MW01377586

PRAISE FOR *THE EVOLUTION OF FAITH*

"*The Evolution of Faith*, the third in a trilogy dealing with religion and science co-authored by Thomas McFaul and Al Brunsting, is an astute exploration of how the discoveries of modern science can interact with the insights of religious pluralism to shape how Christians view God and Christ. The authors develop a new perspective called 'evolutionary pluralism,' and their cross-disciplinary collaboration promises to stretch readers' minds and ultimately clarify and strengthen faith."

—ALYCE M. MCKENZIE, Director of the Perkins Center for Preaching Excellence, Perkins School of Theology, Dallas

"In a world where social media and politics drive us to extremist positions, it is refreshing to see a book that takes a historical look at religious and scientific beliefs and provides a detailed analysis of how they can be—and are being—reconciled. The authors bring science and religion together in a way that upholds and strengthens the core tenets of each. This book is recommended for anyone who loves science and God and wants to understand both more deeply."

—GRAHAM WILLS, Data and Artificial Intelligence Architect and Inventor

"The third book by Thomas McFaul and Al Brunsting, *The Evolution of Faith*, provides a new approach to the discussion of science and religion. They use plain language that is approachable to most readers. Their new insight, called 'evolutionary pluralism,' links evolution and peaceful relations between science and people of different faiths who live close to each other. Their contribution is both unique and valuable."

—JAMES NELSON, former Director of the Asia Technology Center at ITW

"The first eleven chapters of this book lay out the impact of major scientific advancements on our understanding of the universe and provide a view of the various religious beliefs that have grown up in different cultures throughout human history. These chapters provide the groundwork for the powerful, insightful, and thought-provoking final chapter on the evolution of faith. I recommend this book for readers, regardless of their religious or philosophical beliefs."

—DALE ZIMMERMAN, Attorney, recipient of the 1989 Milton Gordon Award for Outstanding Service

"The authors are experts in their fields of science, world religions, and evolutionary history, with long careers examining and teaching these subjects. An excellent review is given of the main leaders in evolving new scientific discoveries, and it is well illustrated. This is an exceptional book for college students, seminary students, and life-long learners to explore humanity's ultimate questions."

—RICHARD L SCHWENK, Cornell University, former Director of the Rural Training Center in Malaysia

"This book is recommended for people who are interested in learning how the world's various religions compare with the current scientific view of the creation and evolution of the universe and the rise of humanity. In addition, the authors show how a single view of the universe with all its complexities relates to religious pluralism and why this is an important connection for humanity to make."

—HOWARD LANGE, Physicist, SETI project

The Evolution of Faith

The Evolution of Faith

Christ, Science, and World Religions

THOMAS R. McFAUL
AL BRUNSTING

WIPF & STOCK · Eugene, Oregon

Wipf & Stock
An Imprint of Wipf and Stock Publishers
199 W. 8th Ave., Suite 3
Eugene, OR 97401

www.wipfandstock.com

PAPERBACK ISBN: 978-1-6667-0237-8
HARDCOVER ISBN: 978-1-6667-0238-5
EBOOK ISBN: 978-1-6667-0239-2

07/13/21

I, Al Brunsting, dedicate this book to my many teachers and mentors who freely shared their extensive knowledge with me. I especially thank Dr. Harry F. Frissel, Chair of the Hope College Physics Department, Holland, Michigan; Dr. David Marker, Assistant Professor of Physics at Hope College; and Dr. Paul F. Mullaney, Group Leader, Biophysics Instrumentation Group, Los Alamos Scientific Laboratory, Los Alamos, New Mexico.

I, Tom McFaul, dedicate this book to the teachers and colleagues who taught me much and to the many students who taught me more.

Contents

List of Figures

List of Tables

Preface

This is the third book in a trilogy of books that the authors have written on the topic of Science and Religion also called Science and Faith. The first book shows how the finely tuned cosmos increases our confidence that an Intelligent Creator called God brought it into being 13.8 billion years ago and guided its evolution toward conscious, self-aware life on earth.[1] The second book focuses on how random events are an inherent part of our universe, from the micro level of atoms to the macro level of galaxies and how God influences our earthly lives in the presence of all this randomness.[2] This current and third book focuses on how Christians can combine modern science and religious pluralism. As a result of integrating these two distinctive areas, McFaul and Brunsting develop a new Christian perspective on the evolution of faith, which is called Evolutionary Pluralism.

The authors are especially well qualified to address these topics. After receiving his PhD from Boston University, McFaul has taught on numerous campuses and received two outstanding teaching awards. He has also held various administrative positions and published eight books and many articles on comparative religions, ethics and social sciences, philosophy, and futures studies. Brunsting received his PhD from the University of New Mexico and completed graduate work at the National Los Alamos Scientific Laboratory. In addition to teaching at Auburn University, he is a highly accomplished optical physicist and engineer; and he holds many US and foreign patents. He is a past recipient of the Bayer Corporation's highest award for distinguished technical achievement. It is the combination of McFaul's contributions to the humanities and Brunsting's accomplishments in science that makes their partnership unique among all other authors who write on these subjects.

---

1. McFaul and Brunsting, *God Is Here to Stay*, 44–66.
2. McFaul and Brunsting, *God and Randomness*, 71–90.

The twelve chapters in this book are easy to follow. They are woven together seamlessly, and each chapter flows smoothly from one to the other. In chapter 1, we—the coauthors—lay out the topics that appear throughout the remainder of the book. These include the big questions related to the origin and operation of the universe, whether there is a spiritual power greater than the universe, and assuming that such a power exists, how it relates to the universe and humanity. We address these questions from both a scientific and religious viewpoint. We define how science and religion differ as two distinct kinds of human experience and show how they are integrated. We describe the differences between premodern and modern science and place our discussion of religious pluralism within the context of our expanding global village. This sets the stage for the rest of the book.

Chapters 2 through 6 focus on the development of science from the premodern to the modern eras. Chapter 2 centers on our early understanding of the material world as envisioned in the unaided eye image of an inverted bowl. We discuss the contributions that the premodern Greeks made in preparing the way for the transition to modern science. In chapter 3, we focus on the period of transition from the premodern and modern science time periods. We highlight the significance of the Aristotelian and Ptolemaic earth-centered or geocentric view of the universe and how the writings of Nicolaus Copernicus and the aided eye, telescopic observations of Galileo Galilei, and the mathematical calculations of Johannes Kepler laid the foundation for our sun-centered or heliocentric theory of the solar system. Then in chapter 5, we show how the transition to modern science was completed through the non-intuitive discoveries of celebrated scientists like Isaac Newton, Werner Heisenberg, Albert Einstein, Edwin Hubble, and Charles Darwin.

Chapters 2, 3, and 4 prepare the way for chapter 5, in which we compare four separate Christian viewpoints of how to reconcile the differences that exist between the premodern creation story in the book of Genesis with the discoveries of modern science. The four views are Young Earth Creationism, Old Earth Creationism, Intelligent Design Creationism, and Evolutionary Creationism. During the course of our discussion, we consider the role of the Bible, the doctrine of biblical inerrancy, and the theory of evolution and how they relate to these four views. Then in chapter 6, we create an innovative approach, for comparing the four Christian positions in a side-by-side comparison table, using the National Academy of Sciences' guidelines for conducting scientific research. Twenty are from NAS, and twelve more are added. The purpose of this comparison is to determine how consistent each of the four Christian viewpoints is to the standards of modern science.

Then, in chapters 7 through 11, we identify how the discoveries of modern science relate to religious pluralism and shape our view of God. For each of the world religions, we include innovative images that identify how their followers envision the relationship between the spiritual and material levels of human experience. Chapter 7 focuses on both the demographic and normative aspect of religious pluralism. We show how electronic communication and mass transportation are altering our awareness of the diversity of spiritual communities that exist around the world and how the interaction between the followers of diverse faiths is increasing. We describe the two universal elements that appear in all religions: the search for the Ultimate and the Universal Reciprocity Norm or Golden Rule.

Chapter 7 also highlights how the world's religions differ in the ways in which they combine materiality, spirituality, transcendence, and immanence. Building on these distinctions, chapters 8 and 9 compare the dissimilarities that exist between the monotheistic and non-monotheistic faiths. In chapter 10, we detail the three ways in which Christians and the devout members of other faiths can respond to those who hold different beliefs: exclusivism, pluralism, and inclusivism. We also show how the followers of any given spiritual community can interpret their sacred scriptures to support any one of these three alternatives. Chapter 11 pulls together the main findings of the preceding chapters and centers on how Christianity can incorporate the discoveries of modern science (chaps. 2–6) and religious pluralism (chaps. 7–10) in answering the big questions that appear in chapter 1.

In the last chapter, chapter 12, we focus on the evolution of faith by advancing a new position, called Evolutionary Pluralism, that includes the main topics discussed throughout the entire book. Evolutionary Pluralism is an alternative to Young Earth Creationism, Old Earth Creationism, Intelligent Design Creationism, and Evolutionary Creationism. In addition to incorporating the discoveries of modern science and the growing global trend toward increased interaction among the followers of the world's diverse spiritual communities, Evolutionary Pluralism remains solidly within the history and traditions of Christianity by building on the life, ministry, words, and works of Jesus Christ.

Acknowledgments

Many people have read and made substantive recommendations to us while this book was being written. Walt Marcum, an associate minister, Highland Park United Methodist Church, Dallas, Texas, has made many important comments and suggestions about the text, about our basic themes, and about our conclusions. We are deeply indebted to him for his contributions.

Joyce E. Brunsting, Al Brunsting's spouse, has been invaluable in making suggestions that avoid unnecessary jargon and that make this book more accessible to a wide and diverse audience. We thank Jim Nelson, PhD, chemistry, who has substantial experience in product development and technical leadership. Also, Robert T. Lehe, PhD, philosophy, suggested that we closely examine each of the four Christian responses to modern science and evolution. This was very helpful.

Also, these people were particularly helpful throughout the review process. Alyce McKenzie, national opinion leader in homiletics, made several important suggestions. Ric Schwenk, PhD, Cornell University, has over thirty years experience in teaching information technology, leading agricultural development in Malaysia and the Philippines, and he has had substantial engagement with several world religions. Besides being an attorney, Dale Zimmerman is well read in theology and the physical sciences. His critical review is much appreciated. There were many others that helped us along this path.

Any and all shortcomings that appear throughout the pages of this book fall entirely on our shoulders.

Abbreviations

This table is intended to quickly reference an abbreviation in the text for its full definition. That abbreviation can be found in the index to locate a discussion of that term in the text of the book.

Abbreviations	Definition
bya	billion years ago
cmb	cosmic microwave background
DNA	deoxyribonucleic acid—genetic material in living organisms
EC	Evolutionary Creationism
EP	Evolutionary Pluralism
IDC	Intelligent Design Creationism
NAS	National Academy of Sciences
OEC	Old Earth Creationism
QP	Quantum Physics
YEC	Young Earth Creationism

1

The Big Questions

During June 2016, we, the authors and our spouses, went on a land tour of some of Alaska's amazing natural wonders. Like the multitude of visitors who travel to Alaska every year, we included in our tour an excursion to Denali National Park. The park derives its name from its central focus—the overwhelming and beautiful mountain called Denali that is also known as Mt. McKinley. The word Denali comes from the native Athabaskan language and means "the high one." At 20,310 feet above sea level (6,190 meters), it is the tallest mountain in North America.

Given the recurring overcast weather of the area, the mountain is cloudy 80 percent of the time. On June 14, we became members of what the locals call the 20 percent club, because during our short stay at the park on that sixty-degree day with light winds, there was not a single cloud in the sky. Visibility seemed infinite, and we saw Denali is all of its glory. On our bus ride to the base of the mountain, we could briefly glimpse a partial view of the mountain in the distance; but this did not prepare us for what we witnessed and experienced at the base of the mountain.

Once there, we saw that Denali's wrinkly conical shape was covered with snow and ice down to about twelve thousand feet. The mountain stood alone and seemed to cover about a third of the whole sky. No other peaks or alpine features competed for the mountain's awesome magnificence. Sunlight reflected diffusely off the snowfields and glaciers that contrasted with the blue, late-morning sky and green tundra on the small rounded hills nearer to us. We lingered there for as long as we could, soaking in all that splendid beauty. In past years, our travels have taken us to some of the most majestic natural wonders the world has to offer, such as the Colorado Rockies, the California Sierra Nevada, the Canadian Rockies, and the Swiss Alps; but Denali tops them all. See Figure 1.

Figure 1. View of Denali, highest peak in North America.

During that brief awe-inspiring encounter with one of our planet's most amazing natural monuments, our minds and emotions raced in two directions at once. The first was scientific. We wondered about the age and origin of the high one that stood before our eyes. How old is it? How did it come to be? How was it formed? What are its physical dimensions? Are there glacier ice fields along its surface? And so on. The second was spiritual. What is the nature of the power that can create something so majestic and beautiful? How are we mortal humans connected to this power that leaves us humbled and inspired at the same time? We lacked words, equations, or materials of the mountain that could capture the spiritual sensation that we were standing in the presence of God the Creator.

We guess that the feelings of amazement that stirred within us at Denali on that bright, sunny day are not unique to us. Year after year, the millions of travelers who are drawn to the natural wonders that exist on every continent bear witness to this double sensation that entwines mind and spirit. To our knowledge, we humans are the only species on earth that is both conscious of its existence and curious about its origin and destiny. We alone ask when and how the cosmos started, where it is going, and where we fit into the grand scheme of things. At our core, we stand apart from other creatures because of our curiosity, which in turn leads us to the next question: how did we develop this capacity?

COGNITIVE REVOLUTION

According to modern scientific theory, modern humans began evolving about 200,000 years ago in eastern Africa. While this might seem like a long time ago, it is a short period compared to when life first appeared on earth about 3.8 billion years ago. To put this time span in another way: Think of the time for life on earth, a life-year, as one year, twelve months, and 365 days. On the life-year time scale humans lived on this planet starting in the last twenty-eight minutes of the day December 31, starting at 11:32 p.m. until midnight of that year, pretty short.

As we evolved during these twenty-eight life-year minutes, especially between thirty thousand and seventy thousand years ago,[1] our thinking and communications advanced. We developed more of a sense of our in-dividuality, our self-worth, our self-awareness, our sense of community, and a sense of wonder. Patterns and predictability became more important. What were the seasonal grazing patterns of the caribou we hunted? If my arrow tips were sharper and more pointed, would they be more effective on the group hunt? This important milestone has been called the Cognitive Revolution. One of the main components of this revolution consists of the development of language that enabled members of various human groups to communicate with each other in symbolic ways. Where are the animals to be hunted? How can we organize the next group hunt? How do we divide up the kill with the group? The capacity for linguistic interaction allowed our ancestors to better protect themselves from predators and to share their views of the patterns of nature that they experienced all around them.

Language also helped people to develop interactively new tools such as spears for hunting and knives for butchering, which in turn contributed to improving their living conditions and lengthen their lifespan. Over time, small wandering hunting and gathering human groups began settling into small self-sustaining villages where agrarian food production supported a

---

1. This is less than the last ten minutes on December 31 on the life-year scale.

growing population of people whose descendants became the inhabitants of the large cities of ancient civilizations.

The development of farming was a huge innovation in human history about ten thousand years ago,[2] or only at 11:58 p.m. and thirty seconds on our life-year scale. The first farmers must have cultivated, watered, and weeded in favor of those plants that they wanted, such as rice and wheat. Communications with other farmers were essential to compare methods and increase their crop yields. This required language skills (including listening skills) and memory from season to season. They must have chased away animals they did not like, such as rodents, birds, and snakes.

Eventually helpful animals were domesticated from selective breeding, such as sheep, cattle, and horses. Those early farmers labored in their fields by clearing the land, planting the seedlings, fertilizing, and watering the growing plants. This was hard, continuous work. Farms required experimentation and recognizing what worked. Eventually, at harvest all their labors, planning, and testing paid off (in some years) with crops that could be used to help feed their families and sustain the members of their communities.

In addition, symbolic forms of communication helped reinforce commonly shared spiritual experiences and beliefs. As a result, there was an increase in the number of priests and religious specialists who became the guardians of a group's religion. Language skills supported stories of God or the gods, which were handed down orally from generation to generation. Over time, as spoken linguistic sounds became transformed into written alphabets and language, many of these stories became recorded and defined as sacred scriptures.

Also, during the early stages of human evolution, curiosity began to emerge, expand, and be reinforced as language and other forms of symbolic communication developed. To ensure their survival, humans directed this growing curiosity toward improving their knowledge of the structure and behavior of the physical world as well as in applying it to enhancing the day-to-day routines of life. They also sought ways to combine their religious beliefs with their understanding of nature as they searched for answers to the big questions related to the origins and destiny of the cosmos and their place in it.

---

2. This occurs in the last one minute and twenty-four seconds on the life-year scale.

FOUR BIG QUESTIONS

The curiosity that sprang forth in the human consciousness as modern humans developed over time parallels the same curiosity that we experienced that June day in 2016 as we beheld the sheer grandeur and beauty of Denali. For as long as we modern humans have walked the earth, we have been astounded by our planet's natural wonders and puzzled over the universe and our place in it. Despite the myriad images and stories that vary from culture to culture about our origin and destiny, at the heart of all of them are four big questions that serve as the catalyst for this book. See Table 1 below.

Table 1. Four Big Questions

1	Where did the material universe come from and how does it operate?
2	Is there a higher spiritual power that is greater than the universe?
3	If so, what is the relationship between this spiritual power and the universe?
4	If a higher spiritual power exists, how does it relate to humanity?

The more we thought about these four big questions, the more we realized that they can be grouped into two major areas that serve as the central focus of this book. The first is science and the second is religion. While each of these areas is distinct, they also interact. Not everyone agrees on the extent of interaction, and opinions vary from keeping them separate to integrating them.[3] Later in this chapter and in chapters 2–5, we will explain how modern science differs from its premodern ancestors in the ways in which it approaches the four big questions. We will also describe how the world's diverse religions respond to them as well. Since the words science and religion carry multiple meanings, we start with defining how we are using them to avoid confusion.

SCIENCE AND RELIGION

What is science? Science is a way of understanding our material world, using evidenced-based rationality. Science primarily focuses on discoveries about our material world, with resulting testable predictions that can be made and compared with proposed explanations (hypotheses). Science is based on measurable evidence and it follows logical and rational steps in the

---

3. For a discussion of the ways in which science and religion are different from each other as well as interconnect, see Barber, *Religion and Science*; Clayton, *Oxford Handbook of Religion and Science*; and McFaul and Brunsting, *God Is Here to Stay*, 1–22.

interpretation of this evidence and in making future predictions. Another characteristic of science is that its explanations are falsifiable, based on verifiable evidence. The word science also refers to the accumulated knowledge, using this process.

Science follows evidence from observations, experimentation, and simulations (based on accepted natural laws). Science is a collection of accepted understandings for how the natural world works. Scientific information is constantly being updated, leading to updates in our understanding of the world. Humility is needed in science. A related characteristic in science is curiosity, which means an openness to new information and pro-actively seeking it out. New verifiable facts are welcomed, even if they do not fit the current worldview.

Here are two examples (of many) that illustrate global contributions of modern-day science. Figure 2 shows the increase in life expectancy averaged for the planet's total population.[4] The scientific contributions that played a role in this amazing increase include advances in vaccines, improved surgical methods, and enhanced communications. Figure 3 shows the level of worldwide hunger.[5] The scientific contributions for this example include advances in genetic engineering of food plants, improved transportation for meat and produce to markets, and refrigeration for food products. These are only two examples of the abundance of scientific contributions that have benefited humankind.

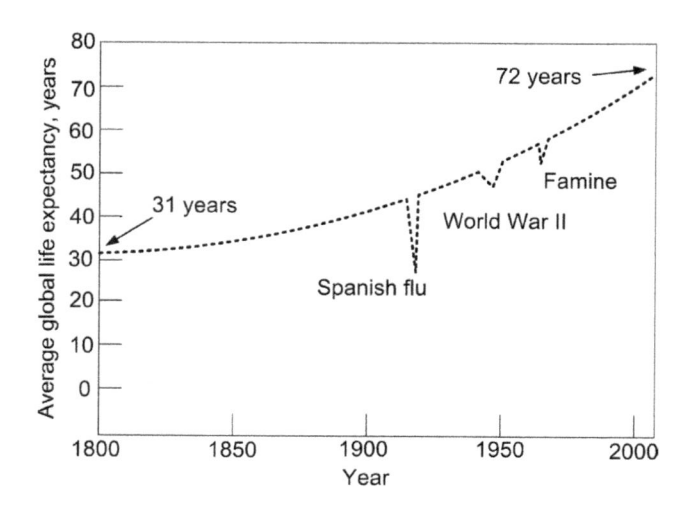

Figure 2. Average global life expectancy from 1800 to 2017.

4. Hans Rosling et al., *Factfulness*, 55.
5. Hans Rosling et al., *Factfulness*, 62.

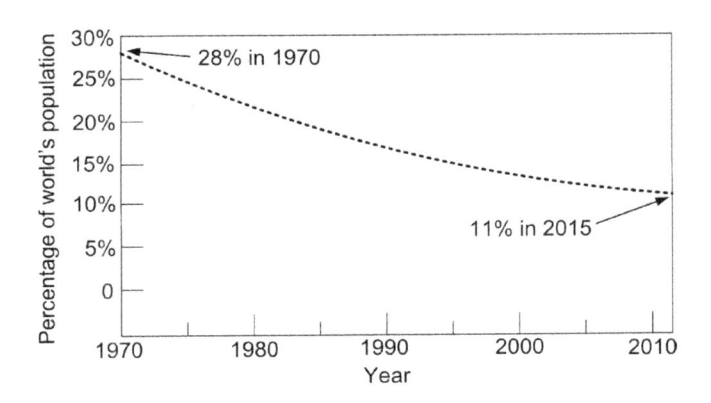

Figure 3. Share of the world's population that are undernourished.

How does religion differ from science? At its core, religion consists of beliefs in a spiritual power or powers that go beyond materiality or that spiritual goals comprise humanity's highest aspirations. This level entails the human quest for meaning or purpose in life and in anchoring it in a shared spiritual vision of ultimate reality. We understand the word spiritual to include both a transcendent and imminent dimension. If a spiritual power stands outside of the physical cosmos, we call it transcendent. If it is imbedded within the physical cosmos, we label it as imminent. Thus, we envision that a supreme spiritual power can be related to the material universe in three possible ways. It can be: (1) both transcendent and imminent, (2) transcendent only, or (3) imminent only.

An example of the first of these three possibilities is Hinduism, which includes the belief that a spiritual power called Brahman transcends the material universe and is also immanent within it. In the case of transcendence, there are two possibilities: with or without intervention. Deists hold that a transcendent spiritual power created the universe but does not intervene within it like the watch maker who makes a watch that runs on its own gears or digital processes. Monotheists such as Jews, Christians, and Muslims perceive that a transcendent God can intervene into history and has done so in shaping the events through prophetic figures such as Moses, Jesus, or Muhammad. For many followers of the modern ecological movement or members of some indigenous religions, a spiritual power is only immanent within or identical with the material universe. We will discuss all three of these possibilities in chapters 7–11.

In addition, we differentiate the concept of spirituality from religion, although they are often used interchangeably. The word spirituality refers to the beliefs different individuals and groups hold to be sacred. They are internal to those who hold them, and they usually are expressed in the form of "I (or we) believe . . . so and so to be true about the ultimate spiritual power of the universe." We understand religion to mean those organized groups that have developed doctrines, scriptures, rituals, and codes of ethics that are an extension of their core spiritual beliefs.

Building on these definitions, we turn our attention to the next critical concern. When we contrast the currently broad scientific consensus about how the universe began, evolved, and is governed by natural laws (chs. 2–5) with the diversity of beliefs that exist among the world religions (chs. 6–10), we are led to wonder why this difference exists. There is one science but many different religious beliefs. Although not everyone agrees, why should there be one dominant scientific view of the origin and development of the material cosmos and multiple religious perspectives on the existence or nonexistence of a spiritual power or powers that transcend materiality and/or is immanent within it? The answer to this question is that science and religion are two distinct areas of human experience and inquiry; and because of this dichotomy, they use different methods to arrive at two types of knowledge. Another way to say this is that science and religion are two separate ways of knowing. They both have different methods, understanding, and thinking.

The following familiar example that is shared by people all over the world whatever their age or station in life illustrates how science and religion differ. When someone holds up an ice cream cone and asks: "Do you see this?" we will answer yes if we are normal-sighted and not severely eye impaired or blind. In other words, we use our five senses to acquire knowledge of our material surroundings. In the case of the ice cream cone, we can see it, taste it, touch it, hear it when the ice cream is scooped out of a container, and smell it when a sweet-scented topping is poured over it. This colorful example is analogous to all the daily encounters that we have with our natural environment. As we go about our everyday routines, we interact with our material universe through our five senses.

While it might appear as an oversimplification, the ice cream cone example is analogous to the world of science. When scientists conduct research to discover how the universe began and evolved, they are working in the realm of the material world. Like the example of the ice cream cone, scientific knowledge starts with research about what we can experience and learn through the five senses. As we will show in chapters 2–5, we have advanced from premodern science where knowledge about the material universe was attained through the unaided eye to modern science where we now acquire knowledge with the aided eye based on accepted modern scientific methods and instrumentation.

For example, modern scientific instruments include the Hubble Space Telescope, atomic force microscopes, the Large Hadron Collider, and positron emission tomography scanners. In addition, modern science has taken us into areas that extend beyond the five senses, such as distortions of space and time due to massive objects (general relativity), quantum effects, constancy of speed of light (special relativity), and the use of complex mathematics to describe and simulate physical processes. As a result, while not everyone agrees, scientists in multiple fields from astronomy to zoology share a broad consensus about how the cosmos started and developed.

In keeping with the ice cream cone example, we turn to the next question. We can ask the person who is holding the ice cream cone, "Do you believe you have a soul?" If the answer is yes, then we can follow up with a second question: "Can you show us your soul?" The response will be no. Why is this? The answer is that the soul is spiritual and not material, which implies that it lies hidden within the physical body and cannot be experienced through sight, sound, smell, taste, or touch like the ice cream cone. Or as the saying goes, "It can't be seen with the naked eye." Nor can the soul be detected by using modern scientific instrumentation and technology. This does not mean that the soul does not exist. It means only that we do not have access to it through science.

Contrasting the tangible ice cream cone with the invisible soul demonstrates that there are two different ways of understanding reality, which is the basis of the distinction between science and religion. The first involves our knowledge of the material world that we start our understanding through the senses and discover through scientific methods and sophisticated research tools. The second centers on the beliefs we hold about whether there exists a spiritual force that differs from the physical universe. Another way to say this is that the human experience of reality consists of two layers. One is material, the other is spiritual.

One of the main differences between atheism and theism is that many atheists believe that only the first layer of materiality is real and that the only form of true knowledge comes from the senses as determined by the discoveries of modern science. Theists do not accept this limitation; they are not reductionists who either reject the spiritual level or attempt to explain it away in terms of materiality. Instead, what theists do is add a second layer of spirituality that goes beyond materiality. Where theists differ among themselves is in how they perceive this second layer and connect it to the first. In chapters 6–9, we will examine how the different faith traditions understand spirituality and relate it to materiality.

In sum, science and religion deal with two distinct levels of reality, the material and the spiritual. Many atheists acknowledge only the first level, while theists accept both. Where theists differ is over whether the realm of the spirit is only transcendent over materiality but not immanent within it, immanent within it but not transcendent over it, or both transcendent and immanent.

This takes us to the next question. How do science and religion and the differences that exist between them relate to the four big questions that are listed in Table 1? To answer this question, we start by dividing the practice of science into two distinct time periods. The first is the era of premodern science, and the second of modern science. The transition from premodern to modern science occurred around 1543 CE when Copernicus first published an audaciously new understanding of the sun-centered universe. Subsequently, this view was confirmed by Galileo's telescopic observations, which were superior to the unaided-eye observations made previously. Prior to this time, premodern scientific knowledge was based on observing the starry skies with the unaided eye and recording the nightly movements of heavenly bodies. After Galileo, modern knowledge of how the universe operates began to emerge from observations and experiments done with the aided eye through the development of increasingly sophisticated scientific tools and methods.

The result of this transition was revolutionary. With the development of modern astronomy, a different image began to take shape as the centuries-old geocentric or earth-centered perception of the universe gave way to a new heliocentric or sun-centered view. In chapters 2–5, we trace how our knowledge of the universe changed from the premodern to the modern era. We describe the contributions that Copernicus, Galileo, and later scientists made to our current understanding how the cosmos came into being, evolved, and continues to expand.

The transition from premodern to modern science changed some of humanity's fundamental thinking. During premodern times, various isolated cultures around the world combined a belief in God (or gods) with the origin of the universe, the destiny of the universe, and humanity's place in it. Premodern science (understanding of the universe and its relation to humanity) and premodern religions were mostly interwoven into a seamless pattern.[6]

After this transition the earth-centered (geocentric) view of the universe was replaced by the sun-centered (heliocentric) view. In addition,

---

6. For excellent references to myths related to the creation and fate of the cosmos, see the following books by Leeming: *Creation Myths* and *World of Myth*. Also see Leeming and Leeming, *Dictionary of Creation Myths*.

these changes altered how we address the four big questions (see Table 1). As modern science emerged, it separated the first question from the other three. The current quest to discover the laws of nature through empirical research does not require a God hypothesis. In a word, the need for God or the gods dropped out. As a result, one of the major changes that arose as modern science replaced premodern science was the separation of science from religion.

This does not mean that God does or does not exist. It means only that it is not necessary for scientists to either believe or not believe in God for science to progress. Nor does it mean that all scientists stopped believing in God, although many did and still do. It means merely that for scientists to conduct research it is not necessary to speculate about spiritual powers that transcend the material world, intervene within it, or are immanent and hidden within it. Modern scientists use empirical and inductive methods and very sophisticated research tools to obtain knowledge about the laws of nature and their applications. If scientists start speculating about the existence or nonexistence of any kind of spiritual power or powers that go beyond the material universe or are connected to it some way, they have left the arena of science and have entered into the field of philosophy or theology.

Likewise, when theologians start by assuming the existence of a higher power and then theorize about its relationship to the material cosmos, whether knowingly or unknowingly they have entered the field of science that seeks to understand how the universe started, how it evolved and is still evolving, and where humanity fits into it. In chapters 2–5, we will explain how modern science transitioned out of premodern science and how the practice of science changed because of this transition. Then, in chapter 6 we will turn our attention to Christianity and describe how Christians differ among themselves in defining the relationship between science and theology. Specifically, we will show how starting with different approaches to biblical interpretation, Christians hold four distinct positions on how to combine the Bible and science.

We are now ready to address our earlier question about why most scientists accept a single scientific view of how the universe began and developed, while the vast majority of the followers of the different world's religions adhere to a diversity of spiritual worldviews. Here is the answer. As we stated earlier, science and religion are two different ways of knowing. Modern science is based on empirical research to discover the laws of nature that govern the material universe. As we will show in chapters 2–5, this has led to (1) a broad consensus about the criteria and procedures that are acceptable for the conduct of modern scientific research, and (2) what constitutes legitimate scientific knowledge that results from it.

When we shift to the field of religion, we observe that there is no consensus on the nature of a spiritual power or authority that might transcend materiality and/or be immanent within it. Different religious communities begin with different spiritual assumptions, which we describe in chapters 6–9. Furthermore, there is no known scientific methodology that we can apply to either affirm or deny the existence of the spiritual images that are associated with different world religions. The result is that in the realm of science, there exists one dominant paradigm of the material universe and its operating natural laws, which emerged over time with the development of modern science; but in religion, there is no dominant view of the existence of a transcendent or immanent spiritual reality beyond nature. Instead, there are many such views that arose in the ancient world and persist to the present day.

RELIGIOUS PLURALISM

Finally, before turning to the next chapter, we address one other issue: religious pluralism. Because of scientific advancements in electronic communication and jet transportation systems, our planet is being transformed into an expanding global village where once isolated cultures are interacting to a degree never experienced in human history. In the realm of religion, the hallmark of this transformation is increased interaction between individuals and communities with dissimilar spiritual beliefs.

High mountains and wide oceans no longer stand as impediments to interfaith contact and communication. Modern scientific innovations in electronic communication and mass transportations systems have accelerated the growth of religious pluralism in the expanding global village. In the last chapter, we examine the role that the Christian faith can play in contributing to greater peace and justice amid the global trend toward greater religious pluralism.

CONCLUSIONS

Before turning to chapter 2, we conclude with the following table (Table 2), which summarizes into four cells the major issues that we have discussed throughout this first chapter. It identifies when the shift from premodern to modern science began, and it divides science and religion into materiality (level 1) and spirituality (level 2) as two distinct ways of knowing.

Table 2. Premodern and Modern Science, Materiality, and Spirituality

Epoch	Level 1, material	Level 2, spiritual
Premodern (before 1543 CE)	Cell 1 Premodern scientific understanding of the universe	Cell 2 Multiple religions emerge in cultural isolation
Modern (after 1543 CE)	Cell 3 Modern scientific understanding of the universe	Cell 4 Multiple religions interact in the global village

As Table 2 shows under the heading of Epoch, the date that separates premodern from modern science is 1543 CE. Under the heading level 1 (material), cells 1 and 3 mark the transition from the earth-centered to the sun-centered view of our solar system, which we describe in chapters 2–5. Under the heading level 2 (spiritual), cells 2 and 4 refer to the world's multiple religions that we compare in chapters 6–10 and that began in isolation centuries ago and now interact with greater frequency in the growing global village. This will set the stage for chapters 11 and 12.

We are now ready for chapter 2.

2

Early Understanding of Our Material World

As we have indicated in chapter 1, we humans have desired to master and understand our physical (or material) world throughout our history. We also have speculated about the linkages between the material and spiritual worlds. The common human pursuit to understand our world is an unbroken bond that connects us to countless preceding generations, and we expect it will continue to do so as the future unfolds. This quest is rooted in the Cognitive Revolution that led to the emergence of modern humans and that separates us from other nonhuman species. During the early stages of modern human evolution, curiosity began to emerge and expand, and this led to improving knowledge of the structure and behavior of the physical world.

PREMODERN DISCOVERIES AND THE UNAIDED EYE—THE INVERTED BOWL

Prior to the emergence of Galileo's telescopic observations in the early seventeenth century, people observed the movement patterns of the heavenly bodies with the unaided eye. They looked skyward, made observations, and compared their results. They thought of possible causes for what they saw to help satisfy their curiosity. For example, one of the earliest explanations for why the stars move as they do was to imagine a huge inverted pottery bowl that covered the sky. Such clay bowls were common during early civilizations. It was necessary to store grain and to protect it from small rodents that would eat and contaminate it. This imagined bowl had tiny holes in the curved sides. It was so large that it covered the whole earth. On the far side of the bowl there were bright lights. The holes of the bowl made these lights appear like points of light in the nighttime sky.

All of the stars appeared to move together from early evening until just before dawn the following morning. This was because the bowl and its holes all moved around the earth together. The bowl needed some kind of prime mover to make this idea work with regularity. So, the points of light, stars, and their movement together during the dark nighttime hours became an accepted explanation for stars and their movements. This early example encapsulates one of the major elements of science: The inverted bowl image (or equivalent) was an imagined way to comprehend almost all of the observations, universally made by people around the world over the years. A sketch of this concept is shown in Figure 4.

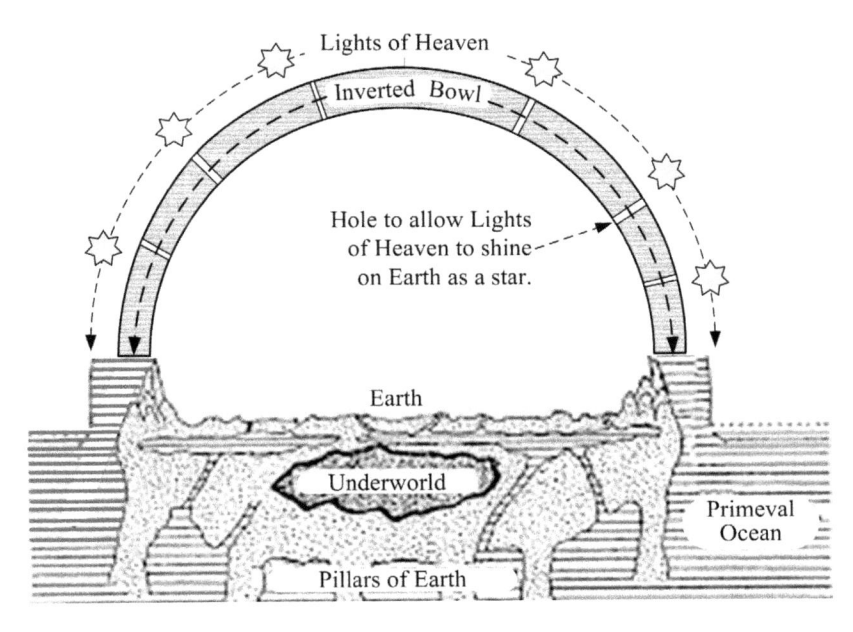

Figure 4. An ancient view of the universe.

The inverted bowl image also captures another key scientific feature: predictability. Based on observations of the nighttime sky and Figure 4, the inverted bowl hypothesis predicts that all points of light in the nighttime sky would move together, and most of them seemed to do that.[1] However, careful observations over weeks and months of time showed that there were a few "stars" that did not move together. These were later known as planets. This meant that the bowl hypothesis (proposed model or theory) did not work for all the points of light. What's going on? Careful record keeping of the positions of the planets ultimately contributed to the transition from premodern to modern, as we show in chapters 3 and 4.

The example of the inverted bowl helps to illustrate many of the common features that apply to both the premodern and modern science eras. (1) Science seeks testable explanations for observations, such as the stars all moving together in the nighttime sky and where the points of light come from. (2) Explanations are expected to be predictable so that the bowl explanation works for each nighttime set of observations. (3) Flaws in the explanations may appear when all of the observations are carefully examined, such as movements of planets. And (4) the bowl hypothesis apparently required some kind of prime mover to move that bowl continuously every day. Note that the bowl revolved around the centrally located humans on the ground, reinforcing the self-centeredness of the earth and *Homo sapiens*.

In addition, the collection of scientific observations is updated constantly. For example, a comet might appear in the nighttime sky that was not seen before. An update would be needed for the previous observations with the new observation of the comet. For an explanation to be tentatively accepted all of the observations at that time must be consistent with that explanation. In this case, the bowl hypothesis would be expected to include a justification for the newly observed comet. If that was not possible, the current explanation would become more doubtful and a new or updated explanation would be sought.

This bowl example applied to the stars illustrates the kinds of analogies (or models) people use to better understand the material world. Ideally, this should be done in an unbiased manner. For example, if the observations of the movements of the planets were ignored so that the bowl explanation would be more believable and less doubtful, then that would introduce a bias that would favor the bowl explanation. This implies that scientific understanding depends on all the observations and not on any preconceived explanation or bias a person might have.

During the premodern science era, the world's diverse cultures combined both the material and spiritual dimensions of human experience. Believing that God or the gods created and influenced the course of natural and human events, premodern people sought to understand how spiritual

---

1. A hypothesis is a proposed explanation based on limited observations.

powers affected their lives. They wondered about how the nonphysical forces could be influenced or even manipulated to improve the chance of living a more comfortable and longer life, as for example in producing an abundant harvest and avoiding famine.

The common assumption was that divine beings were responsible for the movements of heavenly bodies (the bowl). Also, it was assumed that these same divine beings influenced important events that affected human life, such as good weather for crops, falling in love, or engaging in warfare. This was the goal of astrology and why it was coupled with astronomy in the premodern world. Many people consulted astrologers for a better understanding of the future that they believed was controlled by God or the gods that moved the bowl, so that they might better prepare for that future as it unfolded.

THE PREMODERN ERA, ASTROLOGY, AND THE ART OF MAKING WINE

Astrology is based on the belief that divine influences produce the pathways and position of the sun, moon, stars, planets, and comets and that this movement determines natural behavior and human outcomes. For example, wine producers have always depended on the yearly grape yield to contribute to success. According to astrology, the position of heavenly bodies, especially the seasonal elevation of the noontime sun, helped to determine when it was time to harvest grapes. Previous measurements, records, and predictions also aided in establishing the right time of year. It was widely known that precise knowledge of the physical conditions related to an abundant harvest increased the chances for maintaining a profitable livelihood and a comfortable standard of living.

Over the years, vintners built a knowledge base that helped them be successful each season. This included knowing how to prune the grape vines and at what time of the year, how and when to fertilize, and how and when to water the vineyard. When the grapes were ready for harvest, the vintners knew which clusters of grapes to cut from the vine, how to best extract the juice, how to store the fermenting grape juice, and how to contain the finished product, such as in wineskins. This knowledge is related directly to the tangible or material elements of wine production in the premodern world. They could see, hear, taste, touch, and feel these elements.

It was commonly believed that making offerings to God or the gods helped improve the chances of success. In order to increase understanding of the future impact of the divinely created celestial objects, grape growers in the premodern era would turn to the experts—the astrologers—whom they believed would provide them with the best available knowledge of how the movement of the heavenly bodies could enhance their future production and profits. Thus, when seeking answers to the four big questions,[2] through the eyes of the premodern winemaker, the material and spiritual worlds were inseparable.

The connection between the winemakers and astrologers of the premodern science era is but one of many examples of how people who lived in earlier societies around the world integrated the physical and the religious dimensions of their experiences. For the purpose of this book, we will not describe all of the many material-spiritual combinations that existed during the premodern time period. Instead, we will focus on two main questions. What discoveries set the stage for the emergence of modern science? Who made them? While many past civilizations contributed to prescientific knowledge of how the physical world works, one in particular stands out above the rest: ancient Greece. As we show below, it was the Greek philosophers who laid the foundation of modern science through their observations, experiments, and speculations about the origin and operations of the material universe.

PREMODERN SCIENCE AND ANCIENT GREECE

The story of how ancient Greece helped set the stage for the development of modern science starts with seven influential philosophers: Thales (ca. 625–546 BCE), Anaximander (ca. 611–547 BCE), Hippocrates (ca. 460–370 BCE), Eratosthenes (276–194 BCE), Socrates (470–399 BCE), Aristotle (384–322 BCE), and Archimedes (287–212 BCE). For many Greek thinkers the spiritual and natural causes of the events in our lives existed in an overlapping space.[3] At the same time, others began separating the material and the spiritual domains thereby anticipating the practice of modern science.

One of the earliest scientific theories came from Thales. He speculated that appearances to the contrary, the earth actually rests on water and not on a substance more sold. (See "Primeval Ocean" in fig. 4.). It was thought that this water caused earthquakes that shook the northern Mediterranean Basin where he lived. In addition, he may have proposed this concept because it was obvious that all living things require water, such as people, olive trees, and livestock. It turns out that this was the wrong explanation but note that this was a natural explanation that did not depend on mythology or an appeal to the Greek gods. It did not require any priests or sacrifices at

2. See ch. 1.

3. Bauer, *Story of Western Science*, 4–5.

the local temple. This explanation was logical and rational: Starting with the assumption that the earth rests on water, earthquakes, human life, and plant life can be logically explained. And third, this theory provided a universal account as opposed to explanations for random or specific phenomena.

Anaximander, a student of Thales, made some remarkably modern speculations. He was an observer of the universe, even though he did not have access to a telescope as Galileo did over a thousand years later. We will take a closer look at Galileo in the next chapter. Anaximander only had his unaided five senses. He thought that anything that disturbs the balance of nature did not last long. An example would be thunderstorms that disrupted weather patterns. He attempted to understand the movements of celestial bodies in relation to the earth, which led to a new kind of philosophical abstraction. Also, he made important contributions to geography.

In the early Greeks we see some beginnings of applications of mathematics to better understand our world. This helped lay some of the groundwork for modern science. For example, consider Eratosthenes, who was the first person credited to determine the size of the earth employing the modern methods of hypothesis, geometry (mathematics), accurate measurements, and associated predictions. His method is summarized in Figure 5.

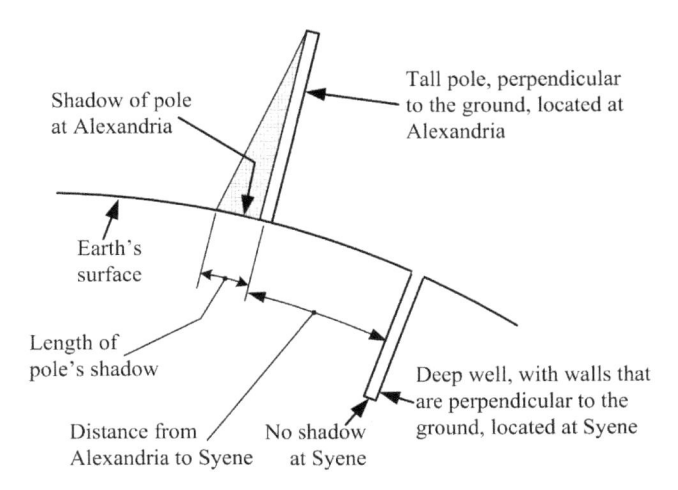

Figure 5. Model of Eratosthenes's method of measurement of the earth's size.

How good was Eratosthenes's conclusion? He calculated that the earth's circumference was fifty times the distance from Alexandria to Syene. He was close. The actual measurement is 47.9 times the distance. He used only the premodern instruments to calculate the length of a shadow in Syene at the same time there was no shadow at the bottom of a well, height of the pole, and distance from Alexandria to Syene. He correctly assumed the earth was spherical (nearly spherical) and that the sun's rays were parallel to each other at the earth's surface.[4] We assume that he applied this result to predicting the shape of the earth's shadow on the moon during a lunar eclipse, a modern application for a new finding. He achieved this remarkably accurate result, given the uncertainties of his measurements. Also, noteworthy was his use of some early and important methods (and thinking) of modern science, such as assumptions, geometry, and measurements.

An early Greek physician, Hippocrates, received an excellent endorsement from Aristotle, who said that Hippocrates was known as The Great Hippocrates. He is recognized as the first physician to contend that diseases were caused naturally. He did not believe that illness was caused by the gods or served as punishment by the gods. Here is another example of an early Greek scientist who omitted religion in favor of focusing on natural forces. Hippocrates's approach was based on the healing power of nature. His idea was that the human body when left alone and under the right conditions can often heal itself. During this time, Hippocrates's contributions to medicine were revolutionary and usually effective. Thinking of him as a premodern scientist, we see an objective observer of illnesses. He must have hypothesized about the natural causes of the maladies that came to his attention, and based on his investigations acted on those hypotheses. This is a remarkably modern approach and is similar to scientific methods used today.

The greatest premodern Greek philosopher was Aristotle. He offered a complete and compelling justification for what was seen in nature so that his works dominated most of premodern Western thought. Aristotle wrote on nearly all disciplines of his day, including biology, logic, and physics. This was in addition to poetry, ethics, and politics. Based on his impressively accurate observations of animals, he provided surprisingly accurate descriptions of their anatomy, their reproduction, and their behaviors. None of these contributions would have been possible without extensive and meticulous observations, recording of notes from those studies, and syntheses and interpretations of this huge amount of information. These are surprisingly modern characteristics of today's science. One of the key differences is that today's scientists use instruments and associated technologies that take premodern observations made with only the five unaided senses far beyond what was possible in those times.

---

4. Given the uncertainties of Eratosthenes measurements the actual fact that the earth is slightly nonspherical and that the sun's rays are slightly nonparallel to each other (on the earth's surface) have no consequence to his determination of the earth's size (circumference).

One of the major differences between Aristotle's thought and modern science was that Aristotle wanted to know what things were in themselves. He wanted to determine the essence of physical objects. What is the plant-ness of plants? What is the water-ness of water? What comprises the celestial objects (sun, moon, stars, planets, and comets) and how are those characteristics different from terrestrial objects? For example, just knowing its measurements (length, width, and depth) of a house would not give any additional insight into the essence of the house. This is why he used five natural classifications (earth, water, air, fire, and ether). Mathematics and precise measurements, which are critically important for many investigations of modern science, were largely irrelevant to Aristotle.

We also see another example for the emergence of modern scientific thinking in the early Greeks with the Socratic method. This approach to thinking is based on interaction between a teacher who asks questions and students who respond interactively. This results in dialogue between students and teacher, which stimulates critical thinking and draws out ideas and underlying assumptions. For example, when discussing politics or ethics, rather than lecturing students Socrates would ask them, "What is justice?" or "What is fair?" He might follow-up with other questions, such as "How can justice be fairly applied to the citizens of Athens?"

The Socratic method encouraged people to think for themselves, use logic, look for unbiased results, interact with each other, listen to each other, and reject inconsistencies. His methodology was not based on accepting the beliefs of outside authorities, such as the mayor of Athens, the priests at the local temple, and holy books. It relied on rational thought and on making logical conclusions that could be compared with those of others. There are many elements of the Socratic methods that apply to the methods of modern science.

The most impressive scientist and technologist among the premodern Greeks was Archimedes.[5] He introduced concepts that were forerunners of calculus that is an important branch of mathematics used in many modern-day technologies and sciences. He used geometry and logic to design and build the tools and effective weapons. He discovered a natural law, Archimedes' principle, that is applied in many ways today, such as in ship design and building. He developed mathematical ways to think about what is now called physics. This conceptually goes beyond only an experiment. He contributed to many disciplines and much of his work is still applied today in our modern world.

For example, in one case, invading Roman soldiers were squeezed into their warships, commanded by Marcus Claudius Marcellus. They sailed

5. Weinberg, *To Explain the World*, 37.

from Rome to attack the Greek city-state of Syracuse (home of Archimedes) on the island of Sicily, which is just southwest of the main Italian peninsula. The Romans expected minor resistance from the undermatched Syracuse defenders. Nevertheless, the invaders were troubled because the Greeks were known to have built effective and technically advanced defensive war machines. Many of these machines were invented and improved by Archimedes. After a string of failures that took two years, the Romans were finally able conquer Syracuse.

A particularly effective Greek weapon, credited to Archimedes, was a grouping of large concave and polished mirrors. From these mirrors the sun's light rays were focused onto an attacking enemy battle ship that wandered too close to shore.[6] As a result the ship caught fire from all that concentrated light and was destroyed. The surprised onboard soldiers in battle armor must have struggled through the sea on their way to shore to save themselves. Not all of them would have made it. Once on shore the defending Greeks must have waited for them with archers, shields, spears, and catapults. To see how this Archimedes mirror system worked see Figure 6.

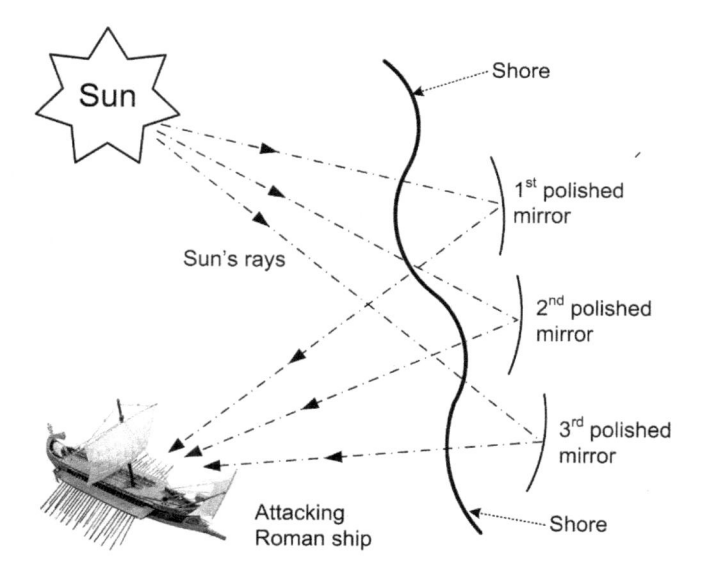

Figure 6. A grouping of large concave mirrors where each one focused the sun's rays of light onto a spot of an attaching warship, causing it to burn and be destroyed. There were likely more than three mirrors as shown here.

6. This story has been pieced together from a number of consistent sources but we do not have conclusive proof of its accuracy.

This arrangement of mirrors is remarkably similar to modern heliostats, used to convert the sun's radiant power to useful electrical power for homes and businesses. Here is another example of similarities between premodern science and technology and modern science and technology, the application of concentrated sunlight for human applications. This Greek weapon required knowledge of how to focus the sun's rays using polished, concave mirrors, how to build and polish the mirrored surfaces, how to arrange the grouping of mirrors, and how to make an effective weapon. All this came from premodern inventors, engineers, and builders who must have used experimentation, optical theory, and mathematics in addition to their engineering and building skills. These are all substantially modern methods. More surprisingly, Archimedes had no modern tools for his calculations such as paper and pencil, computers, and the accumulated knowledge of other mathematicians. Nor was there a printing press.

Another very important premodern scientific example involved determining the value of pi, π, defined as the circumference of a circle divided by its diameter. To four decimal places it is $\pi = 3.1416$. The importance of this ancient Greek achievement continues down to the present day. For example, modern smart phones use radio waves for transmitting and receiving signals. The value of π is essential for the mathematics that describe those radio waves. There are many other applications where π is used, although the long-term significance of this premodern calculation was not fully appreciated at the time it was used. Archimedes devised an elegant and simple method to estimate the value of π, called the method of exhaustion and thereby pioneered the way for many technologies we use today. See Figure 7.

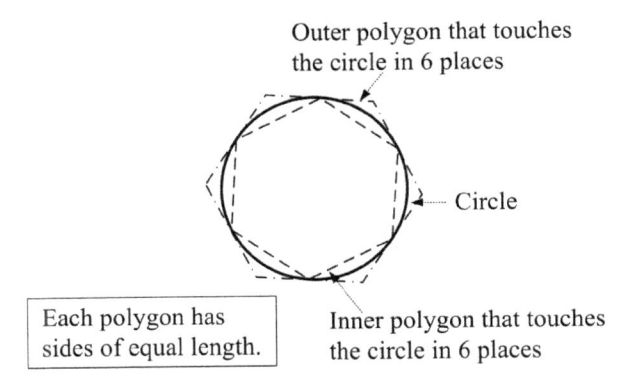

Outer polygon that touches the circle in 6 places

Circle

Each polygon has sides of equal length.

Inner polygon that touches the circle in 6 places

Figure 7. The Method of Exhaustion. The circumference of the circle must be smaller than the circumference of the outer polygon but larger than the circumference of the inner polygon.

Imagine a polygon with six equal sides that inscribe a circle, perfectly touching the circle at six points on its outside. Imagine a second such polygon interior to that circle, also perfectly touching the circle at six interior points. See Figure 7. The circumference of the circle will be less than the circumference of the outer polygon but greater than the circumference of the inner polygon. Calculating the circumference of each polygon is simple and exact.[7] As more sides are added to the interior and exterior polygons their shape will be closer to the circle and the value of π can be more accurately estimated. In the limit of an infinite number of sides the inner polygon will lie on top of the outer polygon and the value of π can be accurately determined. Using this method, Archimedes estimated that the value of π was between 3+10/71 and 3+1/7. In decimals this is between 3.1408 and 3.1429. The actual value (rounded to 4 places) is 3.1416. Archimedes' lower limit was only -0.02% low and his upper limit was only +0.04% high. This is an amazingly accurate result for his time.

This method of Archimedes, the method of exhaustion, demonstrates his advanced thinking and creativity that anticipated the modern branch of mathematics called calculus. Applications of calculus include the determinations of trajectories of rockets that place satellites into orbit around the earth, determinations of water displacement by oceangoing vessels, and determination of the center of gravity for a load of cement blocks that are raised into position by a construction crane. Calculus is extensively used in modern science and engineering for a wide variety of applications. Archimedes certainly helped establish the way forward to modern science and technology with the method of exhaustion.

An important type of practical water pump is attributed to Archimedes, called an Archimedes screw. In many farm fields around the sunny Mediterranean Basin, it was necessary to irrigate the crops with nearby river water. Usually, the level of that water was below the level where it could be used for irrigation. It was critical in those cases to have a means to lift the river water to the irrigation ditches in the farm fields, greatly increasing crop yields. An efficient water pump to do this lifting meant that more of the precious water could be delivered to those ditches for a given amount of energy expended to do that work. The Archimedes screw (fig. 8) proved to be an efficient water pump and it played an important role in successful harvesting of those watered crops. Versions of this type of water pump are used today in many modern applications, such as lifting grain into modern silos, lifting small plastic beads into the input hopper of a modern molding machine, and lifting water

---

7. Using the Pythagorean theorem.

in modern water treatment plants. Here is another example of a premodern invention playing important roles in our modern world.

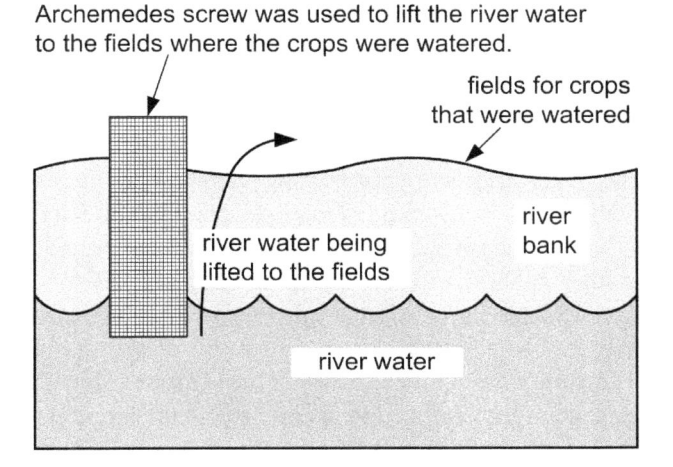

Figure 8. A summary of the Archimedes screw

In another example, King Hiero asked Archimedes to solve an interesting problem that did not involve weapons or water pumps. The king contracted with a local goldsmith to make a specific royal crown to be made from pure gold that the king provided. After several weeks the goldsmith completed the project and brought the crown to the king. Some of the royal advisors wondered if some lower cost silver was substituted for the original gold, which could have been kept by the goldsmith. The king's question was this: How do I know if this beautiful crown contains pure gold as I wanted and not some combination of gold and silver? The only known test for that question was to destroy the crown and analyze the materials, which was not desirable. The king asked Archimedes to determine the purity of gold without destroying the crown (a nondestructive test).

Everyone who takes a bath in a tub of water notices that when you sit down into the water the water level rises and when you stand up in the tub the water level goes down. Surprisingly Archimedes was able to connect these changes in water level with determining nondestructively the purity of the crown's gold. Many times insights into challenging technical problems (like the purity of gold in the crown) are made by connecting two seemingly different concepts. At that moment the great inventor correctly concluded that the amount of water displaced by his body upon entering his bath provided an upward force (or buoyant force) on his body.

Think of a four pound block of wood floating in the water. The level of the water comes up to 10 percent of the top of the block. This displaced water due to the block pushes the block up (against gravity) due to this buoyant force and the block floats on the surface. This is an example of Archimedes' principle. Archimedes was able to invent, design, and build a device to measure the buoyant force on the crown (without destroying it) and determine the purity of its gold. Archimedes was so delighted with this flash of insight he immediately left his bath and all his clothes. He ran naked through the streets of Syracuse proclaiming "Eureka!" or "I have found it." Unfortunately, for the goldsmith, Archimedes was able to determine that the crown was not made of pure gold but of a combination of gold and silver.

In another application of pre-calculus methods, developed by Archimedes, he was able to show that a sphere perfectly embedded within a circular cylinder had a volume of exactly two-thirds of the cylinder. See Figure 9. This contribution is noteworthy because Archimedes did not have these important modern tools: algebra, our Arabic numbering system (1, 2, 3, . . .), or paper and pencil.

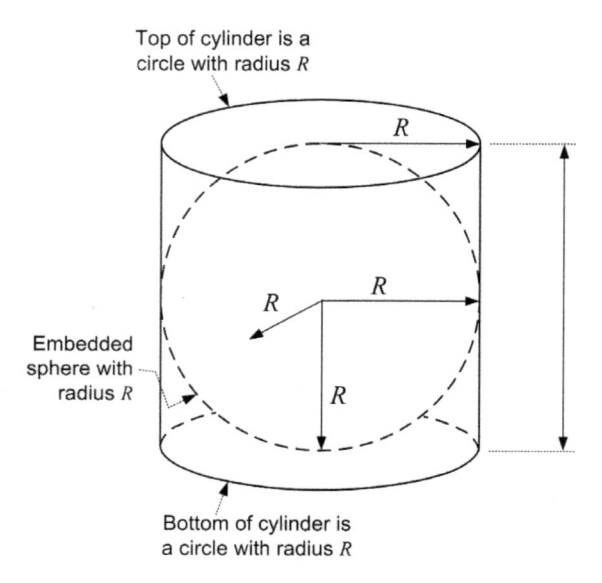

Figure 9. Archimedes was able to show that a perfectly embedded sphere within a circular cylinder had two-thirds the volume of the cylinder.

CONCLUSIONS

The premodern Greek scientists helped lay the foundations for modern scientific methods: They sought natural and general explanations for the world that were not associated with God or the gods. Even though they did not have modern mathematical tools, such as algebra, a numbering system, and calculus, they did apply some mathematics to their proposed hypotheses. They also used some premodern hypotheses to predict other observations and applied them to practical problems that helped improve the standard of living. They made mathematical discoveries that anticipated modern mathematical methods, such as calculus and the value of p. They were driven by their curiosity and desire to know how our material world works; and they assumed that human life could be enhanced through the use of rationality.

From the examples in this chapter we see that premodern science gradually led to the realization that the material world appears to be controlled by natural laws (at least in part). There were regular patterns in the pathways of the stars, linked to divine actions in the heavens (astrology). The size of the earth could be accurately determined from the assumption that the earth was a sphere, light rays traveled along parallel paths from a distant sun, Euclid's geometry, and from a few physical measurements. Also, these natural laws have practical uses, such as determining the gold content of a king's crown. Concerning the four big questions, especially the first one related to the origin and operation of the material universe, human curiosity began to yield results. This Greek premodern scientific legacy helped prepare the way for the eventual development of modern science, which is the focus of the next two chapters.

3

Period of Transition

INTRODUCTION

In this chapter, we describe the period of transition from premodern to modern science. This historic milestone started in 1543 CE and was completed about two centuries later with the work of Isaac Newton, summarized in the next chapter. Before 1543 CE, the earth was thought to be the center of the universe, which seemed obvious to anyone who watched the sun, moon, and stars with their unaided eyes. The stars at night appeared to move in unison from east to west as if they were lighted pinpricks in an inverted nighttime sky that looked like a black bowl as described in the previous chapter.

In addition, in premodern thinking, *Homo sapiens* stood at the center of the cosmos and were the most important beings on earth. Compared to other earthly creatures, humans were more intelligent, we domesticated animals, and possessed the hunting skills necessary for dominating other species. They had language, were self-aware, developed agriculture, and had spiritual sensitivities. This belief was reinforced by the premodern earth-centered or geocentric view of the universe. It was believed that the movement of the stars must have required a divine force. Related to this, people believed that those same divinities influenced (or controlled) their lives, especially in matters of harvest, military battles, and romance.

Building on the achievements of the ancient Greeks as described in the last chapter, the one writer who provided the ancient world with its most comprehensive and systematic premodern scientific view of the universe was the Roman mathematician, astronomer, and geographer Claudius Ptolemy (100–168 CE). Prior to the rise of modern science and the sun-centered or heliocentric view of our solar system, it was Ptolemy who provided the premodern world with a picture of the universe that endured for centuries.

Ptolemy

At the height of the ancient Roman Empire in the second century CE, Ptolemy used geometrical models, based on circles, to explain the observed movement of the sun, moon, stars, and planets around the earth. This model yielded predictions that were as accurate as possible; and when inconsistencies arose, he added epicycles to make adjustments to his predictions. His achievement was of great practical interest to astrologers who were in high demand throughout the premodern world. To achieve this goal, Ptolemy used calculations based on accurate observations of the heavens as a starting place. See Figure 10 below.

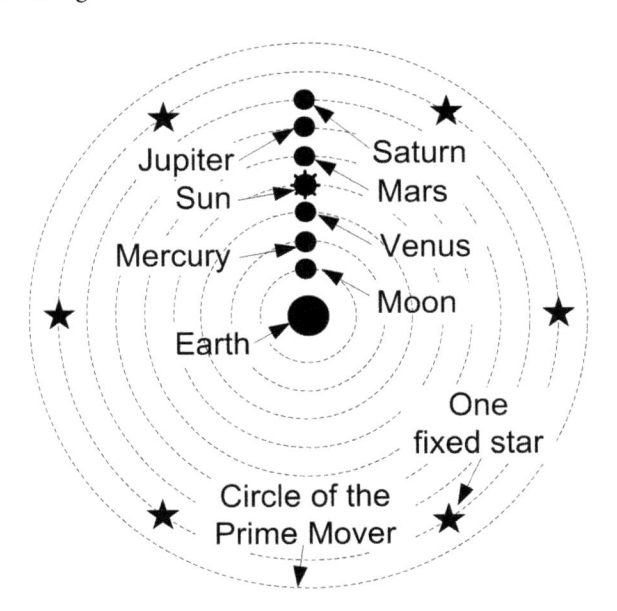

Figure 10. The premodern concept of our universe with the earth as center of the sun, moon, planets, and stars.

The sun, moon, planets and stars all moved in circles around the earth due to a divine Prime Mover, depicted in the outer circle of Figure 10. This is similar to but still an improvement over the inverted bowl concept. See Figure 4 in chapter 2. This concept was mutually consistent with Aristotle's belief that an Unmoved Mover started the universe and with the biblical story of God's divine creation of the universe as described in the first two chapters of Genesis. In elaborating the geocentric theory, Ptolemy used accurate and long-term observations of the celestial bodies from many ancient astronomers. He converted these observations into specific and predictive mathematical results. He also assumed that the movements of the heavenly bodies were along perfect, circular paths. This was consistent with the widely-held Aristotelian assumption that all space beyond the earth was perfect and uncorrupted. Since circles were seen as perfect shapes, all extra-terrestrial objects must move in circles. Here is another example of using a pattern (perfect circles) to explain the unaided-eye observations.

Not everything fit this pattern, however. For example, some planets such as Mars appeared to occasionally move backward relative to the movement of the stars, called retrograde motion. In order to integrate such unexpected retrograde motions into the total pattern, Ptolemy proposed epicycles, or secondary circles, be added along with the primary circles.[1] When he added epicycles that overlapped with the primary circles and filled the gaps, he was able to accurately predict (but not perfectly) the future positions of the stars and planets, as shown in Figure 11. With this addition, the assumed perfect circular patterns above the earth would be saved, making the scheme more widely acceptable in the premodern world.

Ptolemy summarized his results in handy and easy to use tables.[2] Thus, building on the earlier achievements of the ancient Greek philosophers and scientists, he gave the premodern world a finely tuned and universally accepted geocentric mathematical model of the universe. What is particularly remarkable about his accomplishment is that it brought together both the material and spiritual dimensions of human experiences that are identified in cells 1 and 2 that appear in Table 1 in chapter 1.

When Ptolemy's tables became widely used, the church leaders examined his geocentric model, including epicycles. They found near perfect consistency with the Genesis account of creation. Here was a perfect opportunity to integrate Ptolemy's geocentric image with the church's belief that the God

1. Michael Perryman, via Eur. Phys. H37 (2012) 745 arXiv: 1209, 3563 (physics. hist-ph).

2. See Ptolemy's treatise *Almagest*, which was originally entitled the *Mathematical Treatise*. This work was used for over 1,200 years as a guide for predicting the positions and movements of the celestial bodies.

of the Bible was the sovereign Creator of the celestial bodies and that humanity was God's masterwork and centerpiece of creation. In addition, the greatest ancient Greek philosopher, Aristotle, taught the geocentric model where heavenly perfection included circular paths that the heavenly bodies traveled. He held that circles were perfect and that all points on their circumference were the same distance to the circle's center, which was the earth.

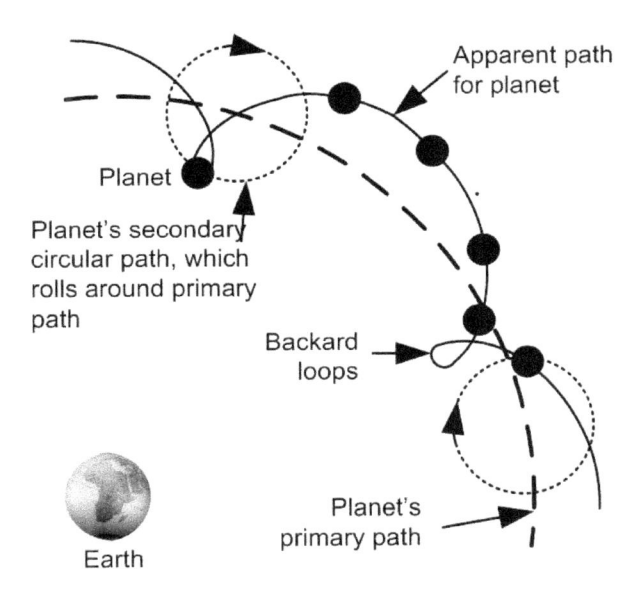

Figure 11. Ptolemy's Geocentric Model of the Universe Using Primary and Secondary Circles or Epicycles.

On top of all this, Ptolemy worked out these wonderfully predictive, easy-to-use tables for the positions of the stars and planets. Any journeyman astrologer could use those tables and prepare horoscopes for his clients. Everyone had an opportunity to see and verify that those predictions were pretty accurate. As the business of such astrologers picked up, other astrologers also gravitated to these tables. Soon most astrologers were using Ptolemy's tables, and the geocentric model became widely accepted and known as the Ptolemaic theory.

If Ptolemy had begun with the premise of a heliocentric model, where the sun is the center of the universe and heavenly bodies moved along noncircular paths, quite certainly his work would not have been accepted. Instead, by starting with the widely established geocentric assumptions of his time, he developed a detailed mathematical model that proved useful in making accurate predictions about the movement and positions of the sun, moon, planets, and stars. To his credit he tweaked the model by adding epicycles or second-ary perfect circles to primary perfect ones in order to improve his predictive precision. As a result, astrologers used his tables to forecast weather patterns, seasonal changes, and how the arrangement of celestial bodies would impact earthly events from wars to when to harvest grapes.

Given the absence of telescopes, satellites, computer-based simula-tions, and spectroscopes to view the heavens,[3] on a clear night anyone could gaze at the stars with the unaided eye and see plainly the inverted, black spherical bowl with pinpricks (stars) that allowed the brilliant light behind the bowl to pass through to the observer on the earth. The mutually sup-portive views of Ptolemy, astrologers, Aristotle, and the Bible, which were reinforced by the pervasive authority of the church, gave the premodern world an earth-centered perception of the universe that was universally ac-cepted and lasted for centuries.

In 1543 CE the premodern understanding of the material world be-gan to crumble as new ideas about the nature of the universe began to take shape. A new model placed the sun at the center of the cosmos and not the earth. Over the next two centuries, several scientists refined this heliocen-tric alternative with a series of new discoveries. In addition to the findings themselves, as a result of the invention of the printing press, they were able to circulate their ideas widely and more rapidly than ever to an increasingly literate public that was eager for new knowledge.

PRINTING PRESS

The significance of the printing press is difficult to overstate and was a key factor in the transition from premodern to modern science. Books, pam-phlets, and tracts could be quickly produced with far less labor, less cost, and in far less time. Prior to the printing press, it was an arduous process to copy every letter on every page of an existing manuscript. Human copying occasionally produced errors that the printing technology avoided. Books from the printing press were exact copies of each other. Also, the earlier copied books were rare, enormously expensive, and owned only by the church, political leaders, or the wealthy.

Before the printing press nearly all books were meticulously handwrit-ten by professional scribes. Typically, one book took months for each copy to be manually produced. Books were rare and enormously pricey, and a scribe's salary was paid from book sales. As a result, most people did not become literate because there were no reading materials they could afford.

3. Spectroscopes are used to measure light intensity over a range of wavelengths.

Print brought the cost per book way down to a level where a much higher percentage of the total population could afford books and other reading materials.

As a result, the production and sale of books after about 1460 exploded across Europe as more people started to purchase and read them. See Figure 12 below. This availability stimulated widespread literacy. The European population that could not afford copied books before the printing press became increasingly exposed to new ideas. In addition, the printing press made it possible for the publication of books, pamphlets, and other materials to be written in the vernacular of common people and not Latin that was the academic and religious language of choice before printing.

Because of the printing press the creeds, rules, and regulations of religious and political authorities could be widely examined and compared with new telescopic observations and their interpretations. This led to challenges to established orthodoxy and facilitated greater acceptance of empirical data and scientific methods. The printing press allowed huge amounts of such information to be distributed quickly, inexpensively, and over wide areas. Eventually books, magazines, posters, product labels, legal documents, and newspapers became readily available to nearly everyone.

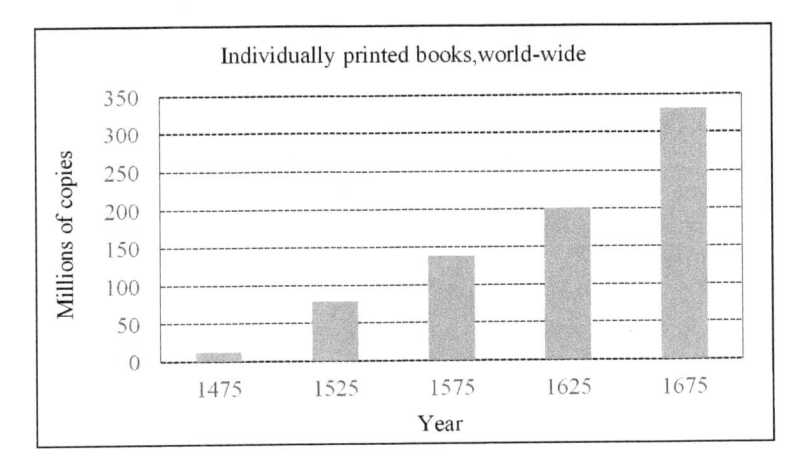

Figure 12. The huge growth of printed books, worldwide.

A printing press is an apparatus that uses ink to transfer text and images from a permanent solid form to individual sheets of paper. The previous technology used handwritten text or woodblocks with letters, designs, and words carved into the wood. Typically, a whole page was carved into wood for each page, which was much more labor intensive, and thus, more costly compared to the new invention. The disadvantages of wood were these: Wood would wear out sooner than the moveable pieces of metal type. Thus, the printed image on each successive page would progressively be degraded due to wearing of the wood. Also, the size of the letters and their shape (i.e., font) could not be changed without laboriously carving a whole new wooden block for the entire page. Gutenberg's moveable metal pieces of type with a single character (for example, a "G") for each piece were used to overcome these prohibitive disadvantages of printing with wood.

Each of the moveable pieces of type (one raised character per piece) was arranged in a frame to be used to print a page. See Figure 13. The pieces were used over and over on successive jobs to print words, sentences, and paragraphs on each page. After the frame was fully assembled with the moveable type, the characters in the frame were evenly inked and pressed onto the blank paper.[4]

Figure 13. Large and ornamental moveable type fills a type case in the Gutenberg Museum in Mainz, Germany.

Gutenberg had to develop the molds (his own invention) to form the metal pieces of moveable type. The press itself had to be adopted to printing from winemaker's screw presses. The ink had to be formulated to give quick

4. A similar technology was invented and used in Korea and in China in the eleventh century, but it never became widely used because each of those languages have a large number of characters, which increased the cost of labor to set the type for each page.

drying, high contrast, no smudging, exact impressions, and last for years. There needed to be storage areas for the moveable type and related necessities in the print shop. The methods to break down the printing forms and reuse it for the next job had to be worked out. Workers had to be hired and trained. The paper had to be compatible with the ink and production process, Printing jobs had to come into the shop so that the new business was economically sustainable. It took Gutenberg several years to successfully work out the whole process so that his printing business became successful.

Figure 14. A page from one of the first Gutenberg Bibles. Notice the detailed decorations, the left and right justifications, and the absence of smudges.

Starting around 1450, Gutenberg started to charge payment for the printing jobs. The first products were simple, such as single announcement sheets and small books. This allowed him to continue to improve his process. Also, this initial work functioned as test cases for his first really major project, printing the whole Bible in 1454. See Figure 14 for an example of just one page from his first production run of 180 Bibles. Throughout each Bible the text was clearly printed, decorated, easy to read, identical to the other Bibles (unlike the manually written versions), and without smudges. This product was obviously far superior to what was previously available at much greater cost. These books were first displayed at a trade fair in Frankfurt, Germany, in 1454. The attendees were astonished with them.

The invention of the printing press facilitated the transition from pre-modern to modern science. For about twelve hundred years in astronomy the Ptolemaic, earth-centered model of the universe was widely accepted in Western culture. It took a really special person to point humanity in a new direction that over time would prove to be more accurate. The new sun-centered theory was originated in great detail by Nicolaus Copernicus (1473–1543 CE). He revolutionized our understanding of the physical world; and in the process initiated the rise of modern science.[5]

COPERNICUS

As a young student during the early days of the European Renaissance, Nicolaus Copernicus enrolled at the University of Krakow, where he studied mathematics and astrology. He was also exposed to the field of astronomy that was closely associated with astrology. He had access to instruments that were used for making astronomical observations and measurements, including a celestial globe that showed the position of the stars. In addition, he had other devices such as the navigational tools that sea captains used for maritime travel. His exposure to all these instruments and the knowledge that he derived from their use helped him develop a detailed understanding of the starry sky. All of these instruments depended on the unaided eye, he did not have a telescope.

Upon completing his studies in Krakow, he entered the University of Padua, in Italy, where he chose medicine as his new field of concentration.

5. Before the rise of modern science, there were other kinds of science besides astronomy, such as medicine, agriculture, metallurgy, domestication of animals, development of calendars. In this book we will concentrate on astronomy, with the recognition that the sciences include other branches. They all have the goal of better understanding the material world and in many cases applying these understandings to the improvement of the standard of living.

By the end of his medical studies, he had earned an outstanding reputation as a competent and well-trained young physician. Upon returning to Poland at the age of thirty, he became the personal physician to an influential bishop while maintaining a strong interest in astronomy. On clear nights, he would venture out of doors to observe and document the positions of the stars and planets as they traversed across the sky.

In all likelihood, the combination of interest in medicine and astronomy reinforced within him a strongly empirical or factual orientation to the study of both the human body and the starry night. As Copernicus's data-driven orientation developed, he gave attention to details. He was a physician, which required attention to medical details. He also collected vast amounts of information about the movements of the planets and stars. In addition to his personal observations, he compiled and shared research from many other astronomers. This is another characteristic of modern science: Relevant observations and data are shared among active scientists.

As a physician the most important results were to cure his patients, even if using novel protocols went against commonly accepted practices. Likewise, as an astronomer Copernicus started to wonder about the result of comparing observations with the prevailing theory. As he continued to accumulate more data, he gradually began to wonder if the geocentric model provided the best explanation for all this data that he had been collecting through his own observations as well as from those of others. It seems likely that his skills as a physician reinforced his skills as an astronomer, such as his attention to detail, his willingness to propose an unconventional diagnosis, and his orientation to verify his diagnosis.

As the process of accumulating additional data continued, Copernicus wondered if there might be a simpler, more elegant theory that would explain better these additional observations and knowledge. In the field of science, it is a common practice that when competing theories exist, the preferred theory is the one that explains the most data in the simplest terms. Simplicity is frequently associated with elegance, both of which are desirable. This preference is called Occam's Razor. With his extensive background in both medicine and astronomy, as well as the growing complexity of the earth-centered model, Copernicus started searching for an alternative and simpler theory.

In addition, by 1543 CE, other astronomers needed better agreement between the geocentric theory and the increasing number of observations that they were documenting. In order to continue their support for the geocentric model, they tweaked the model as they accumulated more data. We have already described how Ptolemy added secondary circular motions to all the planets that were believed moved around the earth. They offset the centers of some of the circles to improve the agreement between observations and the geocentric theory. Keeping circles was important because it maintained consistency with the prevailing traditional explanations. It is well known that adding tweaks to explanations or theories tends to improve the agreement with the observations.[6] This is exactly what happened to the geocentric model. The agreement between data and theory improved at the cost of the theory becoming increasingly complex.

In one of those seminal moments in human history, Copernicus realized that he could explain the growing accumulation of astronomical data by using a simpler sun-centered, heliocentric, theory. This mental shift was nothing short of revolutionary. He had taken the first step toward the development of a new model that opened the door for later scientists whose scientific research confirmed the new Copernican paradigm. See Figure 15. This model is simpler, compared to the Ptolemaic model with epicycles.

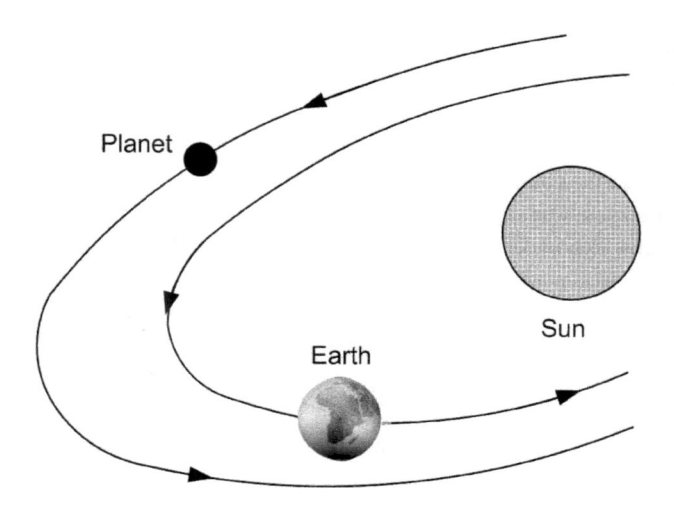

Figure 15. This simple heliocentric model was proposed by Copernicus and eventually replaced the more complicated Ptolemaic geocentric model.

The replacement of an earth-centered belief about the universe by a sun-centered alternative was the beginning of such a major shift in our understanding of how we view the universe and our place in it that we define 1543 CE as the boundary between premodern and modern science. Before this date we thought that the earth was the center of all

---

6. This is another way of adding independent variables, or adding the number of degrees of freedom, to the theory, and thereby increasing the agreement between the proposed theory and the data.

the heavenly bodies, such as the sun, moon, planets, comets, and stars. Copernicus paved the way for an entirely new way of thinking that later scientists confirmed with greater accuracy and elaboration as a result of ongoing experiments and discoveries.

Copernicus assumed that there would be enormous resistance to his explanation by both religious and secular authorities who held that the earth was the center of the universe and that humanity was the crown jewel of creation. He was right. Placing the sun at the center required a radical challenge to the earth-centered theory that Aristotle, the Bible, Ptolemy, astrology, the masses of people, and generations of church leaders had accepted and reinforced for centuries. Just before he died, Copernicus published his heliocentric model and its explanation of the data into a world not ready to accept it. The more the heliocentric model gained credibility in the following century, the more religious authorities resisted it by appealing to the long-standing geocentric views. After Copernicus's death, it was only be a matter of time before the sun-centered model supplanted the earth-centered view as a result of further study and observations.

DEDUCTIVE AND INDUCTIVE REASONING

In addition, a shift away from traditional methods of deductive reasoning and toward newer inductive approaches began to emerge. Like the shift from the geocentric to the heliocentric model, this change was also nothing short of revolutionary. Deductive reasoning was a premodern legacy inherited from many of the ancient Greek scientists. This form of reasoning starts with general statements and moves to specific conclusions. Here is an example:

> General deductive statement: All heavier-than-air matter falls toward the center of the universe.
>
> A less general statement: The earth is made of heavier-than-air matter.
>
> Another less general and related statement: The earth is not falling.
>
> Deductive conclusion from these three general statements: The earth must already be at the center of the universe.[7]

7. Bauer, *Story of Western Science*, 57–58.

By contrast, inductive reasoning occurs when specific (not general) premises are viewed as supplying some evidence for the general (not specific) truth of a conclusion. Here is an example:

> Specific premise: The sun has risen every day so far in my experience, weather permitting.

> General conclusion, taken as truth: The sun will rise tomorrow, weather permitting.

The fact that nobody has ever observed the sun not rising does not guarantee (with complete certainty) that the sun will rise tomorrow (weather permitting). Using inductive reasoning, we may conclude only that it is highly probable that the sun will rise tomorrow, weather permitting. Modern scientists frequently use this method of reasoning to propose explanations (general conclusions) for their observations and experiments, also called hypotheses.

Prior to the emergence of modern science and the development of inductive methods of empirical research, deductions about the nature of the cosmos were authority driven. Also, observations that were inconsistent with the long-held geocentric view were either minimized or dismissed altogether. Copernicus was well aware of this reality during his lifetime. Nonetheless, when he published his heliocentric theory just prior to his death, he opened a new door for others to enter. Using inductive methods that involved the use of his newly invented telescope, it was Galileo Galilei (1564–1642) who took the next big step into the world of modern science.

GALILEO GALILEI

During the time of Copernicus, scientists were beginning to realize that a more accurate understanding of our world required alternative research procedures. This included demonstrations (later called experiments) and mathematically based models. Scientists started to ask how an object behaved, which stimulated more observations and experimentation that led to the accumulation of empirical data. Hypotheses were subsequently developed to explain the data without reference to traditional deductive methods or appealing to existing authorities. Galileo was one of the earliest modern scientists to adopt this approach. According to Albert Einstein, "The discovery and use of scientific reasoning by Galileo were some of the most important achievements in the history of human thought and marks the real beginning of physics."

The data included such quantities as distance, volume, heat, weight, velocity, and acceleration. These observations and experiments required accurate measurements to guide scientists to acceptable hypotheses and better understanding of the world we live in. This demand was met by invention and improvements in known instrumentation. Enhancements came from careful workmanship and were driven in part by merchants who needed accurate measurements of the goods and commodities they bought and sold. Motivation for new instruments frequently came from unmet needs in the marketplace. Also, creative inventors recognized unmet commercial or scientific requirements and successfully designed, constructed, and demonstrated instruments. Galileo made important early contributions to all of this and some of his instruments led to new and critical data and to a new way we see ourselves in this world.

Here are some of the instruments invented during Galileo's time, 1600–1650 CE: telescope, microscope, barometer, thermometer, and the pendulum clock. These instruments and other improvements made it possible to see or to measure quantities that before this time could only be estimated or were beyond the reach of the unaided senses. Many of these instruments advanced modern science, having been developed as a result of newly discovered principles.

As a boy, Galileo had a strong curiosity and was very good at determining how things worked. This led him to his early inventions, for example, his improvements of the early telescope. When he was thirteen years old, his parents sent him away from his home in Florence, Italy, to study at a monastery. As a result, he wanted to become a monk but was discouraged by his father because monks had no income. Likely Galileo was well grounded at this time in Christianity and in church matters, which he used later, as we will see.

He was seventeen in 1581, and moved to Pisa, where he started his studies at the University of Pisa, fifty-three miles away from Florence. He returned to Florence only during the summer to see his family. His father encouraged him to study medicine, which he did. At that time, much of medicine was not based on verifiable scientific data, but on the theories of ancient authorities, such as Hippocrates. Galileo, like his father, tended to question the conclusions that most others took for granted.

People from throughout Europe, such as Copernicus, traveled to Italy to study. As the Renaissance developed,[8] many early modern thinkers and scientists began challenging the traditional views and the authorities who

---

8. The Renaissance included a revitalization of European art and literature under the influence of classical approaches, during 1300–1500 CE.

supported them. It must have been exciting for Galileo to meet and engage students from various cultures with different experiences, different thinking, and different views of science and religion.

Galileo developed a strong interest in mathematics at this time. Medicine became less important to him. Mathematics do not depend on appealing to external authorities for validation. For example, Euclid's plane geometry started with a small number of intuitively appealing axioms, such as the whole is greater than any one of its multiple parts. Other theorems were derived from a given set of axioms, for example, the three interior angles for any plane triangle always add up to 180 degrees. The axioms and subsequent methods of Euclid's plane geometry did not depend on outside authority.

According to Galileo, "Philosophy is written in this grand book the universe, which stands continually open to our gaze. But the book cannot be understood unless one first learns to comprehend the language and read the alphabet in which it is composed. It is written in the language of mathematics."[9] Galileo's used mathematics for much of his work. He also had an underlying desire for the universe to be logical, internally consistent, repeatable, discoverable, and based on induction and not deduction. For example, at the university of Pisa he was nicknamed a "wrangler," which referred to someone who loved to engage in skepticism, dispute, and confrontation. He seems to have been a "wrangler" throughout his entire life.

In 1589, Galileo became professor of mathematics at the University of Pisa. Even though it was not a well-paid position, he enjoyed the company of other faculty members and their diverse ideas. While at Pisa, he became more interested on how objects fell to the ground. While this may seem uninteresting, most people thought that heavier objects fell faster than light ones, as Aristotle had taught. Everyone believed that a lighter feather fell slower than a metal ball.

Galileo developed a simple experiment to test this notion. To avoid most of the effects due to the feather's air resistance, Galileo used two balls. One was much heavier than the other. He climbed to the top of the Leaning Tower of Pisa and simultaneously dropped both of them. To the astonishment of the onlookers both balls struck the ground together, contrary to the accepted ideas of the day. As Galileo and others adopted and increased their use of inductive methods to study nature, their interest in using deductive reasoning faded.

Also, telescopes were invented shortly before Galileo built his own telescope, with which he began to observe the heavens. At that time,

9. Steele, *Galileo*, 23.

telescopes used glass lenses ground to specific shapes that were sections of a sphere. If the lens was thicker in the middle, compared to the edge, it was called convex. This lens is frequently used to magnify the printed words on a page or to focus the sun's rays into a spot of intense brightness. These early telescopes used a convex lens and a concave lens (thicker on the lens edges, compared to its center). These two lenses were placed in a tube to keep their spacing and positions correct and to block out stray light. When done carefully and properly, distant objects appeared closer when viewing them through the convex lens. With careful design the lenses could be used to improve the magnification of a distant object, making it appear closer, compared to the unaided-eye observation.

In October 1608, a Dutchman, Hans Lippershey, gave the first demonstration of such a telescope, which drew widespread attention across Europe. By this time Galileo had well-developed skills for invention, improvement, instrument making, and demonstrating scientific instrumentation. To his credit, when Galileo became aware of the basic design of this first telescope, he immediately used his considerable skills to build an improved telescope of this own. He developed enhanced methods to grind and polish the finest clear glass to make the lenses. During this process, Galileo significantly improved the design, so that a round object that was magnified by three diameters in a previous telescope was magnified by 20 diameters in his improved telescope.

During Galileo's lifetime, Venice was becoming a wealthy city-state, mostly based on trade from around the Mediterranean Basin that came by sea. The city leaders were concerned about alien ships that might damage their economy. It was important to identify vessels before they came too close to the Venetian docks. Galileo's telescope allowed them to do this. When he demonstrated his newly improved telescope to the government officials, they were thoroughly delighted and rewarded him handsomely. Additionally, this new telescope could be put to good use on the battlefield.

Galileo had other ideas and pointed his telescope toward the stars. His newly improved telescope greatly upgraded observations made with the unaided eye. What would he find? Would Aristotle, Ptolemy, and the church leaders who defended the geocentric model of the universe be right? Or would he find something different despite centuries of authoritative certainties based on repeated observations made with the unaided eye? Given his reputation as a wrangler, he no doubt enjoyed making these kinds of observations.

Using his improved telescope, Galileo's observations led to considerable support for heliocentricity and how we understand our place in the universe. He saw the rough and uneven surface of the moon contrary to

Aristotle, who stated that the moon was perfectly smooth, a view that Ptolemy, the church, and virtually everyone else accepted. Galileo made the first ever observations of the moons of Jupiter in January 1610. His multiple observations over weeks showed that the moons were orbiting Jupiter. This was in opposition to the geocentric theory, where all the heavenly bodies were believed to orbit Earth. Here was evidence that the earth was not the center around which the moons of Jupiter orbited. By the end of 1611, Galileo had measured the orbital times for each of the four moons of Jupiter that he had discovered. Table 3 gives Galileo's results compared with modern values in days (d), hours (h), and minutes (m). The accuracy of Galileo's measurements must have meant that he made accurate observations with precise timekeeping and carefully documented results.

Table 3. Galileo's measurement of orbital times for each of the four moons of Jupiter[10]

Name of moon	Galileo's measurement	Modern measurement
Io	1d 18h 30m	1d 18h 29m
Europa	3d 13h 20m	3d 13h 18m
Ganymede	7d 4h 0m	7d 4h 0m
Callisto	16d 18h 0m	16d 18h 5m

Also, Galileo observed the phases of Venus, which are similar to the phases of earth's moon, clearly visible with the unaided eye. Phases of our moon are easily understood as the illumination of the moon over half of its surface by the Sun. By extension, that understanding would also apply to the phases of Venus. This would, however, imply that Venus orbits the Sun and not Earth, in violation of the geocentric model. All of his observations either supported the heliocentric model or did not contradict it. By contrast, his observations did contradict the geocentric model. It became increasingly easy for Galileo to support the heliocentric model.

Using the printing press in 1610, Galileo published these preliminary telescopic observations in a readable book, entitled *The Messenger of the Stars*. The first edition quickly sold out. This illustrated book convinced many of heliocentricity. As this book became more widely distributed around Europe, Copernican heliocentricity became more widely accepted. As this theory, based on Galileo's data-centered telescopic observations,

10. Weinberg, *To Explain the World*, 175–80.

became more dominant, support for the geocentric model of Aristotle, Ptolemy, and the church diminished. For centuries, Christian leaders stood solidly behind the earth-centered model. As the heliocentric view, based on inductive experimentation, gained credibility at the expense of deductive reasoning, they threw the weight of their authority against it, and in particular, against Galileo.

In 1616, the pope used the church's feared court, the Inquisition, to warn Galileo not to promote the Copernican heliocentric model. In response, in 1632, Galileo published a second book, *Dialogue of the Two Principal Systems of the World*, in which he reiterated his support for the heliocentric model. Shortly thereafter, the church banned sales of this second book and ordered Galileo to go to Rome and face the Inquisition. Because he was now elderly and in poor health and feared the real threat of torture, on June 22, 1633, he promised the judges to give up such beliefs. After the trial, the church placed him under house arrest. Then, as his physical stamina diminished, he became depressed. He died in 1644.

KEPLER

Despite the church's personal condemnation of Galileo and suppression of his discoveries, acceptance of heliocentricity continued to gain momentum, especially as a result of the telescopic observations that the German astronomer and mathematician Johannes Kepler (1571–1630) conducted. Throughout his life, Kepler was a stickler for detail and accuracy. For example, he calculated the instant of his conception by his parents as May 17, 1570, at 4:37 in the morning. This was the day after his parents were married. He did this in order to show that he was born prematurely and, therefore, that he was legitimate. His life was turbulent; he was never wealthy, and he frequently had money problems. Several of his children died in their youth. On one occasion later in life, his career was interrupted by the Thirty Years War, and he defended his unconventional mother against the charge of witchcraft in a trial that could have easily sent her to be burned at the stake. He died on November 15, 1630.

When Kepler was young, he lived and worked hard in the inn of his grandfather. Johannes waited on tables, cleaned the eating and sleeping areas, washed dishes, and did whatever else he was told to do. He was a physically limited child. He was near sighted, frail, and not especially coordinated. When he broke dishes in the inn, he was rebuked sternly. However, he was smart and curious about the world around him. Consequently, it was determined that Johannes career choices were limited to the ministry. Similar to Copernicus, Kepler was trained in theology, and he was a religious man.

In 1587, he received a scholarship when he entered Tubingen University. He enrolled in theology and philosophy and studied astronomy and mathematics in order to better understand God's creation. In spite of his focus on theology and philosophy, including the church's support for the geocentric model, he concluded during his time at Tubingen that the Copernican heliocentric model was superior. Upon graduation, Johannes intended to join Tubingen University's theology faculty, but he received and accepted an offer for a post in mathematics and astronomy at the Protestant school in Graz, Austria. He was twenty-two at the time.

Some five years later, in 1598, a Catholic archduke who controlled Graz ordered all Protestant Lutherans to leave. This forced Kepler and his wife of one year (both were Lutheran) to forsake their home. As a result, he came to work for Tycho Brahe, a wealthy astronomer in Prague. A year and a half later Brahe died; and Kepler, whose skills were well recognized, took Tycho's place. This allowed Kepler to focus completely on astronomy and associated mathematical descriptions of his observations.

Building on his mathematical skill, he hypothesized that the planets moved around the sun in elliptical orbits, a major departure from Aristotelian requirement for perfect circles. This was his first of three laws. His mathematical descriptions of planetary orbits brought the Copernican heliocentric model into much closer agreement with astronomical observations and, thus, into nearly complete acceptance. Until that time, the main advantage of the heliocentric model over the Ptolemaic geocentric model was that the heliocentric model was simpler and more elegant, not more accurate. Now, as a result of Kepler's first law, described below, the Copernican model more accurately agreed with current observations. In modern science, if there are two theories and the first is simpler, more elegant, and agrees better with the data compared to a completing second theory, then the first theory is accepted while the second is rejected.

Kepler's main contributions to astronomy and modern science were his three laws of planetary motion. His first law was that each planet, including earth, goes around the sun in a curve called an ellipse. See Figure 16.

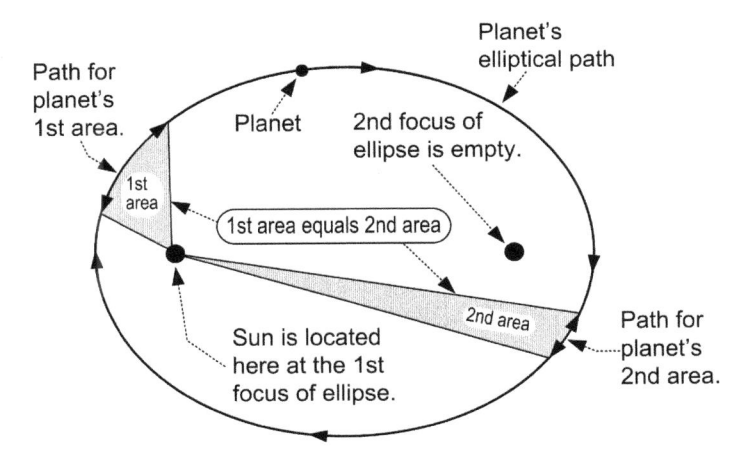

Figure 16. Planetary Elliptical Paths around the Sun.

An ellipse is similar to a circle, except that the ellipse is elongated. The time for a planet to travel on the path of a planet's first area is the same as the time for a planet to travel on the path of the planet's second area. There is a mathematical relation between the amount of elongation and the location of the two foci (labeled as 1st and 2nd focus in fig. 16). The closer the two foci are together, the more the ellipse looks like a circle. The planetary paths known at the time of Kepler were close to being circles. So, it is not too surprising that the Ptolemaic theory, including circular epicycles, approximately agreed with the data. But now with Kepler's first law the heliocentric model the agreement was better. Also, the elliptical orbits are an obvious departure from circles as required by the Aristotelian and pre-scientific geocentric model. Kepler would have known this from his studies of philosophy at Tubingen University.

Kepler's second law is that planets do not move around the sun with a uniform speed. He observed that their speed is faster when they are closer to the sun compared to when they are farther away. As the planet moves, it sweeps out an area defined by the planet's path and a line between the planet and the sun as shown in Figure 16. If the planet is closer to the sun as shown in the upper left of Figure 16, it would sweep out the shaded area (labeled "1st area"). Later when the planet is farther from the sun and traveling more slowly, it would sweep out an area as shown in the bottom right of Figure 16 (labeled "2nd area"). If we assume that those two areas are equal (1st area equals the 2nd area), then the time it takes for a planet to move on its elliptical path around the external arcs of each of the two areas is equal. This law was also supported Kepler's observational data.

Kepler's third law is different from the first two because this law relates the distance that each planet is from the Sun (orbital size) to the time each planet takes to complete one orbit (its period). This law is clearly demonstrated when the distance between a planet and the Sun is given in terms of the distance from Earth to the Sun being a distance of 1.000. Also, for clarity, set the time period for one revolution by a planet in terms of Earth's time period of 1.000. This law says that if we square the time period and cube the distance, the answers will be equal. For example, the distance between Mars and the Sun is 1.524 and its period is 1.881; 1.524 cubed (1.524 times 1.524 times 1.524) is 3.54; 1.881 squared (1.881 times 1.881) is 3.54. Likewise, this relation holds for the other known planets at the time. In Table 4, Kepler's third law applies to actual observations as summarized below.[11]

Table 4. Kepler's third law of planetary motion shows surprising accuracy relating a planet's distance from the sun to its time to complete one orbit around the sun.

Planet	Distance from the Sun*	Time to complete one orbit (years)*	Distance cubed	Time squared
Mercury	0.387	0.241	0.058	0.058
Venus	0.723	0.615	0.378	0.378
Earth	1.000	1.000	1.000	1.000
Mars	1.524	1.881	3.54	3.54
Jupiter	5.203	11.863	141	141
Saturn	9.537	29.448	867	867

* referenced to Earth

CONCLUSIONS

The transition from premodern science to modern science developed over about two centuries and started around 1543 CE. Before the changes that Copernicus and Galileo initiated, anyone could see that the sun, moon, planets and stars made their daily trek around the earth. The earth dominated

11. Love, *Kepler and the Universe*, 161.

the cosmos at its center, and humans dominated the earth. People sought the advice of astrologers before they made the important decisions that determined their destiny. Foretelling of the future was linked to specific movements of the celestial objects that the gods moved, and the gods influenced terrestrial events.

Ptolemy played a key role in the premodern world. His easy-to-use handbook of the motions of the sun, moon, stars, and planets made the earth-centered (geocentric) universe consistent with how the gods influenced the daily lives of people. The Ptolemaic theory enjoyed the support of the church and was bolstered by the reputation of the greatest of the ancient Greek philosophers, Aristotle. Combined with the story of creation, appearing in the first chapters of the biblical book of Genesis, the internally consistent, premodern synthesis supported the earth-centered view of the cosmos.

In 1543 CE, Copernicus was the first to publish a contrary and revolutionary explanation. This date is the boundary between premodern and modern science. His heliocentric, sun-centered, idea postulated that all the planets orbited the Sun, not Earth. Galileo was the second key player to contribute to the passage of premodern to modern science by observing inductively through a newly invented telescope the moons of Jupiter, sunspots, and the phases of Venus. His contributions also included applying mathematics more extensively to his observations and measurements.

Before Kepler, it was believed that motions of the planets were circular, which was based on Aristotle's authoritative writings. Kepler suggested that those paths were elliptical, not circular (his first law of planetary motion). Analysis of the data at that time agreed more closely with Kepler. His second and third laws were also mathematical, which continued to point the way for mathematically based laws of nature. Newton was needed to complete the transition from premodern to modern science, and we meet him in the next chapter, along with others whose amazing discoveries contributed to our current comprehension of the cosmos.

4

Transition to Modern
Science Is Completed

INTRODUCTION

As we described in the last chapter, with the aid of the printing press, Copernicus was the first to propose the radical and counterintuitive idea that the sun, not the earth, was the center of the universe. Subsequent telescopic observations by Galileo all pointed to a Copernican heliocentric universe. Likewise, because of the printing press, these results were well distributed around Europe. As more people read about it, the tide began to turn toward the sun-centered universe. The corresponding transition to modern science received a big boost when Kepler showed that the movement of the planets was explained better if their paths were elliptical and not circular, as was assumed at the time. In this chapter, we demonstrate how the transition to modern science became complete.

ISAAC NEWTON AND CLASSICAL PHYSICS

Building on the work of Copernicus, Galileo, and Kepler, it was Isaac Newton (1642–1727 CE) who took the decisive step that confirmed the sun-centered theory. In 1687, Newton published the *Principia*, in which he explained his three laws of motion and his universal law of gravitation.

These natural laws explained why the planets moved the way they did and they explained the underlying validity of Kepler's three laws of planetary motions. Newton's work was so thorough, so mathematical, and so consistent with experiments and observations, it became the basis of human understanding of the movements of not only the stars but of also of all motion here on earth. As a result, he has been called the "prince of astronomers and philosophers,"[1] and his contributions continue to influence ongoing achievements of modern science.

For example, human-made satellites orbit the earth in accordance with Newton's three laws of motion and Newton's universal law of gravitation. These satellites were lifted into space by rocket-thrusts based on Newton's third law of motion: For every action there is an equal and opposite reaction. In space, the force of gravity that attracts these orbiting satellites to the earth is the same gravitational force that influences our everyday lives. More than anyone who preceded him, his laws, proofs, and insights have provided foundations for modern mathematics, physics, and astronomy.

The premodern scientific view was binary. On the one hand, consistent with our everyday experiences, earth was viewed as a place of uncleanliness, rust, decay, and death. On the other hand, it was believed that perfection existed above the earth in extraterrestrial space where circles were perfect, and there was no decay. Completely contrary to this binary premodern view, Newton began considering gravitation from a different perspective. In his own words: "I began to think of gravity extending to the orb of the moon . . . I deduced that the forces which keep the planets in the orbs must be reciprocally as the square of their distances from the center about which they revolve: and thereby compared the force required to keep the moon in her orb with the force of gravity at the surface of the earth and them [the comparison] . . . nearly . . . equal."[2]

He imagined that gravity experienced here on earth was the same force that kept the moon in its orbit around the earth, completely contrary to the binary premodern view. He proposed a mathematical description for gravity that eventually became known as the universal law of gravitation. See Figure 17 and related explanation. He objectively compared measurements of gravitational forces here on earth with what he expected at the moon and found them to be the same, using the newly discovered law.

---

1. Cohen, *Franklin and Newton*, 318.

2. Newman, *Science and Sensibility*, 123. The reason for the small residual disagreement was that Newton did not have an accurate distance from the center of the moon to the center of the earth.

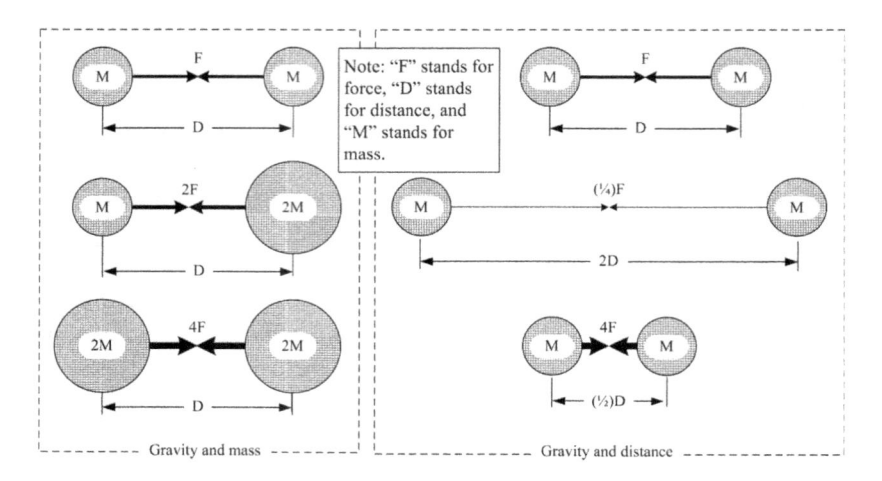

Figure 17. Newton's Universal Law of Gravitation.
Left side, gravity and mass: The mass of each object is either M or 2M, but the
distance between them, D, stays the same. The gravitational force changes from F
(top) to twice F or 2F (middle) and to four times F or 4F (bottom).
Right side, gravity and distance: The distance changes from D (top) to twice D or
2D (middle) and one-half D or (1/2)D (bottom) while the mass, M, stays the same.
The gravitational force, F, changes from F (top) to one fourth F or (1/4)F (middle)
and to four times F or 4F (bottom). Halving the distance, D, to (1/2)D increases the
gravitational force fourfold to 4F.

Before Newton there was no accepted explanation for what held the
planets in their elliptical orbits around the sun, according to Kepler's first
law. Kepler wrongly supposed that these motions were "magnetically" related. Why was the orbital speed of the planet faster when it was closer to the
sun, according the Kepler's second law? Why is the cube of the distance between a planet and the sun proportional to the square of the planets period
or orbit time (Kepler's third law)? The underlaying explanation for Kepler's
three laws of planetary motion was Newton's universal law of gravitation,
which is the basis for explaining all three of Kepler's laws.

Newton suggested that gravity was a force that influences and changes
the motion of a body (such as an ice hockey puck, car, or jet aircraft). That
same gravitational force that kept Earth in its elliptical orbit around the Sun,
the Moon in its elliptical orbit around Earth, and Mars in its elliptical orbit around the Sun also affected everything on Earth. Apples fell from trees,
stones dropped from a leaning tower fell to the ground, and arrows, propelled
by archers with bows, had trajectories that brought them back to earth. This
was an enormously unifying insight: The gravitational force beyond the earth

was the same force that we sense here on Earth's surface. This force was universal and it was mathematically revealed for the first time by Newton.

The universal law of gravitation is also remarkable for its generality, because it applies here on Earth, in our solar system, in our galaxy, and in our whole universe. Planets orbit the Sun, gases move around black holes, man-made space vehicles sent to Mars, and other planets in our solar system all move in accordance with Newton's universal law of gravitation. In fact, every object in the cosmos attracts every other object in the cosmos, an astounding generalization. This law applies not only for space beyond Earth. Based on Newton's gravitation law and his three laws of motion, Kepler's three laws of planetary motion were now explained.

But by the time Newton completed his life's work, the heliocentric model had totally superseded the geocentric theory. We now had answers to these questions: (1) What was at the center of the known universe? Answer: The Sun. (2) Did the planets and moons orbit in perfect circles at constant speed as was taught by Aristotle? Answer: No, in ellipses at variable speeds. (3) What explains the planets orbital behaviors (e.g., times, speeding-up and slowing down, orbital times in relation to orbital distances)? Answer: Universal law of gravitation and Newton's three laws of motion.

Before Newton the movements of objects here on Earth were poorly understood, such as cannon balls, pendulums, and raindrops. Here was another indication that the universe was not binary: earthly and heavenly. Newton's three laws of motion allowed for analysis of movement in both spaces. This was another remarkable generalization by Newton. His three laws of motion are summarized in figures 18, 19, and 20.

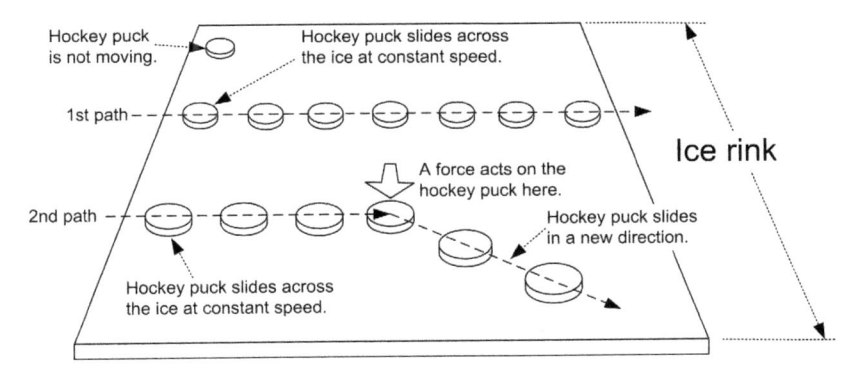

Figure 18. Newton's 1st law of motion. Every object remains at rest (top, puck is at rest and not moving) or in uniform motion in a straight line (middle, puck slides across fiction-free ice at constant speed, ignoring the small frictional force) unless it is acted upon by an unbalanced force (bottom, puck changes direction due to a force).

Figure 19. Newton's 2nd law of motion. Force is equal to mass times acceleration. The more force (torque) the car's motor applies to the car, the greater the acceleration (ignoring friction). If the motor were to apply the same force to a heavier car (more mass) the acceleration would be reduced (all other things stay the same).

Figure 20. Newton's 3rd law of motion. For every action there is an equal and opposite reaction. The backward actions of the jet's exhaust are equal to the force moving the jet forward.

Newton also extended inductive reasoning, initiated in the modern science era to these four rules. (1) Simpler explanations for what we observe in nature are more likely to be correct than complex ones. Occam's Razor. We first encountered this rule with Copernicus. (2) Natural effects that occur in different places, such as apples falling in the United States and apples falling in England will probably have the same causes. (3) If a property, such as Newton's second law of motion (force equals mass times acceleration) is demonstrable on all objects subject to experimentation (e.g., ice hockey pucks, cars, and jet planes), it is likely that property applies to all bodies in the universe. (4) A general law (e.g., Newton's third law of motion, for every action there is an equal and opposite reaction) that is confirmed by trustworthy experimentation or observations, is likely to be true unless additional experimentation or observations make another law more likely.

By the end of Newton's life, it was widely accepted that the modern scientific approach for understanding the laws of nature should be based on observations and experimentation and usually described mathematically. As a result of Newton's work, especially as detailed in his *Principia*, science moved entirely into the modern era. Newton was at least partially recognized in his lifetime for his extraordinary achievements. In 1703, he was elected president of the Royal Society; and in 1705, he was knighted for his scientific contributions by Queen Ann of Great Britain. Since then, his work was recognized much more extensively with worldwide acclaim.

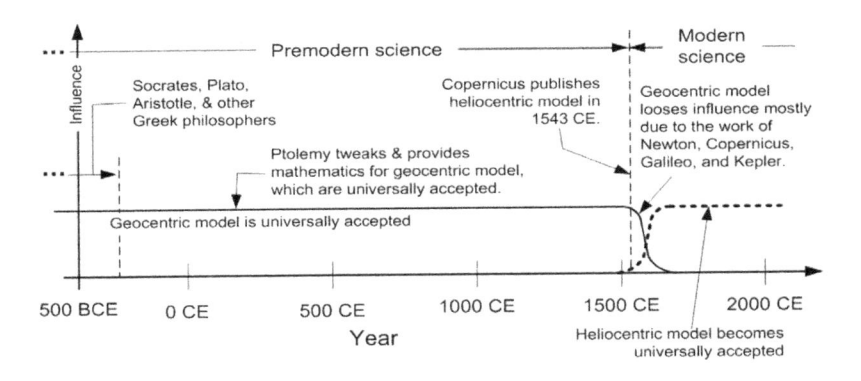

Figure 21. Timeline showing premodern science and modern science.

Figure 21 summarizes the transition from the premodern to the modern scientific time periods. It dates back to the time of the geocentric model that the ancient Greek philosophers and the Ptolemaic theory supported. Starting with Copernicus in 1543, a new heliocentric theory began to emerge; and as a result of the follow-up discoveries of Galileo, Kepler, and finally Newton, the transition to the heliocentric view of the universe was complete. In the next section we see how Newton's classical physics was enlarged to include some very unexpected results that were shown to occur in the material world.

NEW QUANTUM PHYSICS

Let us leave size scales of the very large, such as stars, galaxies, and the universe, and consider the very small, such as atoms and their subatomic particles. There were shocking and historic discoveries to be made at these small sizes. Before 1911 no one knew what an atom really was. In that year Ernest Rutherford (1871–1931) completed a set of experiments that showed nearly all the mass of gold atoms was condensed into a very small nucleus. This result was unexpected. After this discovery scientists began to understand the atom consisted of far smaller subatomic particles, including a positively charged nucleus and surrounded by negative electrons. This raised a new and fundamental question: What keeps those surrounding electrons from spiraling down into the atomic nucleus, losing their energy in the process, and causing an atomic catastrophe? Clearly atoms were stable. A clue came in experimental evidence showing that atoms absorbed light (or more properly radiation) only at certain wavelengths. Also, atoms only radiated light at certain wavelengths. What determines those individual (or quantized) wavelengths? Why wasn't there a continuum of wavelengths for the observed light?

The highly counterintuitive answers to these and other related fundamental questions were worked out from 1900 to 1930, and resulted in quantum physics. Here is one of the basic ideas: Waves and particles seem to be completely different. Light as displayed in rainbows and prisms behaves like waves with shorter, more energetic blue waves. Longer red waves are less energetic. Particles, like baseballs, are individual, stand-alone, clumps of mass. Particles do not spread out along wave fronts. Imagine a baseball batter trying to hit a baseball wave-front with the resulting droplets spraying all around home plate. No, particles are really different from waves. However, it was discovered through experimentation that sometimes light unexpectedly behaves like particles when in rainbows light behaves like waves. Also, it was found through experimentation that particles, for example electrons, can sometimes behave like waves. Waves are sometimes like particles and particles are sometimes like waves. This new understanding was not at all anticipated, using Newtonian physics. A new understanding of reality was emerging that was entirely shocking.

It turned out the new and successful answers (quantum physics, QP) showed why atoms were stable and did not catastrophically implode. It showed why only discrete wavelengths of light were emitted from (or absorbed by) a given collection of identical atoms. QP also explained the existence of the elements of the periodic table (e.g., oxygen, carbon, hydrogen, lead, silver, and gold) and how those elements combine to form stable molecules (e.g., water, carbon dioxide, and ammonia). QP was spectacularly successful and has been experimentally found to work in all applicable situations. For example, QP is essential in the design of our smart phones and related devices. We need QP to understand why the sun shines. QP is needed to understand how heavy atoms (that comprise our bodies, clothes and shelter) are produced in supernovae. This human triumph of scientific theory has widespread application for our understanding of our world, the very small atomic world, and the very large cosmic world. QP was "the most radical innovation in physical theory since the work of Newton."[3]

This abundance of success, however, came at a high cost. QP is not at all reasonable or logical. It has very unexpected consequences. The wavelike character of matter and energy was understood in terms of waves of probability, i.e., the likelihood of finding the associated particle in a small region of space during a short period of time. These probability waves seemed to oscillate in free space and not in a physical medium. Particles themselves were now understood to have only a range of knowable positions and another associated range of knowable velocities, referred to as the Heisenberg uncertainty principle. Contrary to implications from Newtonian physics, you cannot know the position and, simultaneously, the velocity (more properly the momentum) of a particle with perfect accuracy. There is a knowable limit. Fortunately for baseball those kinds of limits are fall smaller than what a batter needs to know how to successfully hit a pitched baseball.

Predictions from QP are used throughout our technical economies. Those predictions are essential for product development, especially our digital electronics like HD television monitors, smart phones, and navigation systems. Yet QP is so counterintuitive and, in some ways, tends to obscure the reality of the very small. In the next section we move to other important discoveries in modern science that were counterintuitive and unexpected in the contributions of Albert Einstein.

3. Weinberg, *To Explain the World*, 261.

ALBERT EINSTEIN

In December 1999, Albert Einstein (1879–1955) was recognized as *Time* magazine's person of the century. He is considered, along with Newton, as one of the greatest scientists of all time. Here we summarize some of his contributions and how they relate to the themes in this book.

Around 1900, scientists thought that nearly all the laws of nature had been discovered and that a complete understanding of the universe was at hand. Only a few minor details needed to be cleaned up. At that time one assumption was that the wave motion of starlight needed some kind of medium to travel to the earth, called ether. This seemed obvious. This is similar to sound waves needing the medium of air for the train whistle to travel from the train to your ear. Once the properties of ether were understood, it was thought we would have a nearly complete understanding of our universe. All the important scientific investigations would have been done.

However, there was one experiment that contradicted this understanding of ether, the Michelson and Morley experiment in 1887.[4] These results showed that the speed of light does not change for the earth moving #1 in the direction along its orbit around the sun, compared to #2 a direction perpendicular to #1. The experimental errors were small enough that detection of such differences in the speed of light was possible. Here is another example of how an experimental result led to a major and completely unexpected understanding of our material world.

This result created a problem. Light was expected to travel at a constant speed through a substance, called the ether, just as sound waves travel at a constant speed through the air. Apparently, the expected analogy between light and ether compared with sound and air was not valid, based on the Michelson-Morley results. If you move in the same direction as light, you would expect the measured velocity of light to decrease, compared to moving opposite to the direction of light where it would increase. Follow-up experiments confirmed the Michelson-Morley results. It appeared that a complete understanding of the universe, especially light, was elusive.

This discrepancy between what was expected and what was observed for the measured speed of light, opened the door for a young Albert Einstein. In a 1905 paper, he proposed a radically counterintuitive solution to this problem. He started with the clearly obvious observation that we cannot tell if we are moving through the ether. It is not detectable. His conclusion was that ether was not needed and, in fact, does not exist, since there is no direct experimental evidence for it.

---

4. Feynman et al., *Feynman Lectures on Physics*, 1:15.3–15.5.

With ether rejected, he assumed that all experimental results are the same comparing no movement with moving at a constant velocity. For example, think of yourself seated comfortably in a stationary train car. You close your eyes and daydream for ten minutes. When you open your eyes and look out the window you see that now the train is slowly and smoothly moving. You do not feel the train moving. From the back of the train, you throw a soft baseball to your friend at the front of the train. She catches it and throws it back to you at the same speed. The speed of the baseball does not change in either direction due to the train's speed. This is an example of Einstein's assumption and in this case it was intuitive.

Now, here is the nonintuitive, nonobvious, master step: Since there was no evidence for the speed of light changing, regardless of the speed of the observer, let us audaciously assume that the speed of light does not change, regardless of the speed of the observer. Think of driving from New York to Boston on I-95 at a speed 65 miles per hour. Your driving speed depends on distance covered (miles) per time it took to cover that distance (hours), distance (or space) and time. The speed links both space and time. If you hear an approaching train whistle while driving, the whistle sounds as if it has a higher pitch. If the train whistle is moving away from you, a lower pitch is heard. You would expect the same would apply to light, both are waves and both travel in a medium (air and ether). Well, light unexpectedly does not behave like that.

Einstein worked out the mathematical details and arrived at a startling and unexpected result: There was a direct equivalency between mass and energy, $E = mc^2$, where E is energy, m is mass, and c is the constant speed of light. One of the consequences of this well-known equation is that when the nucleus of a uranium atom splits (fissions) into two nuclei where the total mass of the two resultant nuclei add together to be *less* than the mass of the original uranium nucleus. The leftover mass goes into a huge amount of energy. This is the basis of the vast destructive power of the atomic bomb and of the peaceful use of nuclear power.

One of the consequences of this theory is that time is not separate from space. They are codependent on each other. Together they form a four-dimensional fabric (three space dimensions and a single time dimension).[5] Einstein's theory is called the "special theory of relativity" and today it is completely accepted by modern science.

The counterintuitive predictions of this theory have been verified in countless experiments around the world, in measurements on satellites, and from the results of atomic clocks. This equivalence of mass and energy

5. McFaul and Brunsting, *God and Randomness*, 154–58.

explains why stars can burn for billions of years before they die out. Here is a requirement of accepted scientific theories, like special relativity: Predictions of the theory must be consistent with trustworthy measurements that are repeatable from multiple, independent investigators. Another requirement is that those documented results are peer reviewed by other scientists in the field who are not associated with the results being submitted for publication.

The special theory of relativity did not cover Newton's universal law of gravitation, which said that if the position of the sun was changed instantaneously (in principle) within the solar system the resulting gravitational forces would instantaneously change around the solar system. This meant that gravity "signals" could in principle be sent faster than the speed of light, which is disallowed by special relativity. Another problem was that absolute time was now required but special relativity eliminated absolute time in favor of relativistic time.

Einstein was mindful of these issues, and in 1912 he had a key insight: he imagined the four-dimensional fabric of space-time to be curved. The amount of curvature in these four dimensions depended on the amount of nearby mass and/or energy. This advanced concept was unique in modern science at that time. Objects near the earth's surface like baseballs, apples, and rain drops, as well as heavenly objects like planets, comets, and asteroids, would all move along straight lines in a curved space-time, observed as gravity in our three-dimensional space. Gravity was not seen as a force field but as a result of space-time curvature. The amount of curvature depended on the amount of nearby mass and/or energy. This new interpretation of gravity provided a more fundamental understanding of gravity compared to Newton's universal law of gravitation. It is not that Newton was wrong and Einstein was right. It is about a deeper explanation of gravity, similar to Newton's universal law of gravitation used to explain Kepler's elliptical orbits.

In 1915, Einstein published his paper on general relativity which showed that the effects of gravity were equivalent to space-time being warped by nearby mass and energy. See Figure 22 for a visual representation of this warpage. This paper was very mathematical but still could be tested by direct comparisons to observations. Here is another example of a pioneering view in modern science that the world is governed by natural laws (like special relativity and general relativity) that were for the most part mathematical and more unified than had been imagined in Newton's time.

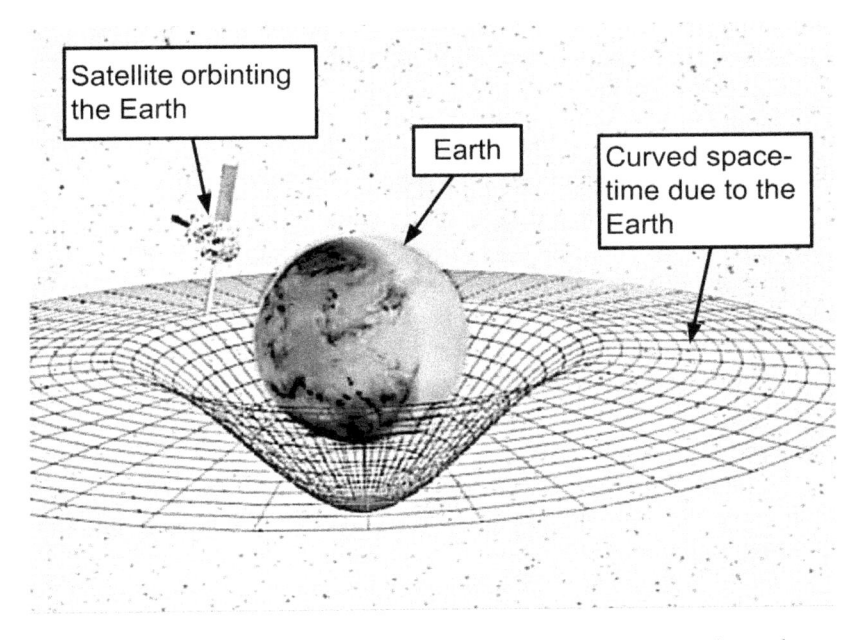

Figure 22. Curvature of space-time holds a satellite in orbit around the earth.

Einstein's theory of general relativity was spectacularly confirmed in 1919. Einstein's equations quantitatively predicted the warpage of space-time near the sun due to its mass and energy. This meant that light rays passing near the sun as observed on earth would be bent. On a British expedition to West Africa, careful measurements were taken during an eclipse. The moon blocked direct sunlight, making that starlight more accessible to the observations. Meticulous analysis of those measurements showed that light from stars that passed close to the sun's rim were bent by exactly the amount predicted by general relativity. Here is another example of an important theory for our understanding of the world that was successfully tested by thorough observational evidence. Since 1919, general relativity has been extensively and successfully tested by other observations.

This new understanding of the warpage of space-time led to a another understanding regarding the big bang, discussed in the next section. When that event occurred 13.8 billion years ago it was the space-time that expanded, carrying along the mass and energy of the universe with it. Therefore, the question of "what is the universe expanding into" has no meaning since the mass, energy, and space-time are expanding.

Next, we turn to new understanding about the size of the universe, which is one of the big questions that we humans have pondered since before the early Greeks of premodern science.

EDWIN HUBBLE, THE BIG BANG, AND
THE EXPANDING UNIVERSE

Based on his theory of general relativity, Einstein concluded that the universe was stable and static, a steady-state universe. This meant that the universe has no beginning and no ending, which was the prevailing view in the early 1920s. Up to this time, astronomers observed fuzzy patches of light using more primitive telescopes, compared to what are used today. The Milky Way galaxy was thought to be the whole universe and the fuzzy patches were viewed as clouds of luminous gas. Like the Copernican heliocentric revolution nearly four hundred years earlier, our understanding of the universe was about to be radically changed.

In 1919, the astronomer Edwin Hubble arrived at Mount Wilson Observatory in Southern California. A new, state-of-the-art 100-inch (2.5 m.) Hooker Telescope for his use awaited him. A critical step in Hubble's research occurred in 1923 when he identified a distinctive type of flickering star, a Cepheid star, within a fuzzy patch called M31. These types of stars were very special because their pulsations are directly related to their luminosity (brightness). The Hooker telescope had high sensitivity and high resolution and so it became possible to determine how far a Cepheid star was from the earth by measuring indirectly its flickering speed and apparent brightness.[6] Using telescopic measurements of this star, Hubble determined that the Cepheid in M31 was at least 930,000 light-years from the earth.[7] Since the Milky Way galaxy is about 100,000 light years across, the fuzzy M31 patch was located beyond our galaxy.

Hubble's observations and measurements led to a stunning conclusion of historic proportions. The universe we inhabit is larger than the Milky Way. At that time, it was thought that the Milky Way was the extent of our universe. Starting with Galileo's first telescopic observations and continuing to those of Hubble's, astronomers were discovering progressively larger size of the universe. None of these discoveries would have been possible without the use of sophisticated telescopes and advanced technology. Also essential

6. The method for this distance measurement was developed by Henrietta Leavitt and Harlow Shapley.

7. A light year is a unit of distance (not time). It is the distance light travels in one year, or about 6 trillion miles or 9.5 trillion kilometers.

were the people who made detailed observations, documented their results, and drew valid conclusions, even when they challenged the current views of the time.

Over the next ten years, using the 100-inch Hooker Telescope, Hubble continued his observations of fuzzy patches called nebulae. After conducting careful research and drawing documented conclusions, he concluded that there were at least forty-four thousand other galaxies scattered throughout the universe. The earth's galactic home, the Milky Way galaxy, was clearly not alone. Almost four hundred years earlier, Copernicus had initiated the correct idea that the earth is one of multiple planets orbiting the sun. Hubble extended this conclusion to another level by observing that there is multiple, indeed, thousands of other galaxies besides the Milky Way.

Hubble made another astonishing discovery. He found that generally the greater the distance a galaxy was from earth the faster it was receding from the earth. "The nebulae [galaxies] are all rushing away from our stellar system [Milky Way], with velocities that increase directly with distances."[8] Contrary to Einstein's theory of the steady-state universe, Hubble established that the universe is changing and expanding. As an analogy, we can envision a partially air-filled balloon with letters on it. If we continue to blow air into the balloon, the letters become enlarged and spread further apart as the balloon expands. Comparing the balloon to the universe, the modern interpretation of this expansion is that the fabric of space-time is expanding, and the galaxies are carried along within that fabric.

This observed expansion, along with other internally consistent evidences, must mean that in the past, the universe and space-time were smaller. Also, in the future, the universe and space-time will be larger. These conclusions ultimately led to today's inference that there was a starting point, a big bang, 13.8 billion years ago.

On January 29, 1931, Einstein and Hubble sat together in the plush leather seats of a Pierce Arrow touring car. A chauffeur drove them up a winding mountain dirt road to Mount Wilson, where the Hooker Telescope was located. Einstein wanted to see how Hubble made the observations and measurements that challenged his own view of the steady-state universe. As a result of Hubble's demonstration, by the end of the trip, Einstein had changed his mind: the universe was expanding. Over the next several years, Hubble continued with his astronomical mapping of the universe. By 1940, most astronomers and physicists agreed that the universe is not static and in a steady state. Instead, it is changing and expanding. This discovery of

8. Hubble, *Realm of Nebulae.*

the expansion of the universe by Hubble "was one of the great intellectual revolutions in the twentieth century."[9]

In the next section we move from cosmology to biology and meet a modern scientist who offered an amazingly general explanation for the diversity of life on earth, Charles Darwin.

Charles Darwin

The remarkably predictive modern theory of evolution was formulated and first published by Charles Darwin (1809–1882) in 1859. Christianity's relationship with evolution and modern science will be discussed in chapter 5. During Darwin's childhood, there were no financial worries. He lived comfortably. At this time, he collected specimens, fished, watched birds, and read about natural history. However, his academic performance was not up to expectations, which concerned his widowed father. Darwin's father decided that Charles, now sixteen, was to become a physician like him. Young Charles was sent to the University of Edinburgh in Scotland, where he enrolled in medical studies. Darwin found that he disliked the sight of blood and, as a result, found his medical studies were repulsive. He focused his interests again on the natural biological world by collecting fossils and learning more about animal life. This new direction concerned his father because he thought his son might become just an idle gentleman without contributing much of value to society.

Taking a new direction, Dr. Darwin enrolled his son in the ministry at Christ's College, University of Cambridge, starting in the fall of 1827. However, Charles developed a heightened interest in the field of naturalism when he attended lectures on botany given by Rev. John Stevens Henslow. Darwin became more enthusiastic about biology, which he saw as a possible career path. Even though he was uninterested in studies for the ministry, Darwin managed to pass his final exams in January 1831, placing a respectable tenth in his class.

Shortly thereafter he was invited to sail on the English survey ship *HMS Beagle* as the naturalist. The assignment was to chart the coastal waters around South America and beyond. The ship's captain was also youthful. Robert Fitzroy was twenty-three years old and Darwin was twenty-two. Fitzroy was interested in evidences that confirmed a literal biblical explanation of creation. Darwin was well suited for this post since he was trained as a clergyman and he developed an expertise in biology. During the voyage, starting on December 27, 1831, it became clear that Darwin did not agree

9. Hawking, *Brief Answers to the Big Questions*, 46.

with biblical literalism, and this became a source of lasting friction between the two men.

The ship's voyage took Darwin southward along South America's eastern coast and around its southern point into the Pacific Ocean. The most important stop for Darwin was in the Galapagos Islands. These islands are tropical and volcanic. They are spread out on either side of the equator in the Pacific and lie about 560 miles west of the South American coast.

Significantly, Darwin found that each island had its own species of finches.[10] Each type was well suited for the kinds of food on its specific island. When Darwin went to a different island with somewhat different kinds of food, suitable for finches, he found a different species of finch that was well adopted for that type of food on that island. For example, finches on one island had heavy, strong beaks that were effective for cracking open the seeds that were readily available on that island. On another island Darwin found another species of finch that was well adopted for grasping and probing, useful for catching insects that were more abundant on that island.

There were finches on the mainland too that were similar to each finch species on a given island. Darwin hypothesized that different finch species on the islands were descended from the ancestor finch species on the mainland. Perhaps, the finches on the islands were originally carried there by winds where they subsequently developed adaptations specific for each island. Modern scientists have analyzed the DNA from these finches and confirmed Darwin's hypothesis was correct. After four years, nine months, and five days, Darwin returned home (October 2, 1836). He never left England again.

While on his voyage Darwin sent many specimens home to England and stowed others aboard ship. There was a huge amount of material for him to study and develop his ideas. While in seclusion, Darwin organized his scientific records and cataloged his specimens. Gradually he started to develop his assumptions that would eventually be published. His thoughts involved four questions, written in his personal notebook: (1) What evidence is there that species are changeable? (2) How do species adapt to a fluctuating environment? (3) How do new species come about? (4) Why do some different species have similarities? In the fall of 1838, Darwin read *Essay on the Principle of Population*, by Thomas Malthus. Implications of this work were that human population will constantly increase and that the increases in food supply will not keep up with the population increases. The conclusion was that human population will constantly be restrained by

10. A species is a group of living organisms (such as the small birds, finches). Within the group there are similar individuals capable of exchanging genes or interbreeding. The species is the principal natural taxonomic unit.

overall food supply. Darwin noted in his journal: "It at once struck me, that under these circumstances favorable variations would tend to be preserved and unfavorable ones to be destroyed."

By the winter of 1838, Darwin, who was very thorough and detailed, had corresponded with experts in various related fields, such as dog breeders, farmers, and veterinarians. This helped him to refine his theories about changes within species and from species to species. From 1837 to 1856 Darwin shared his emerging theories about the origin of species and the supporting evidences with a few confidants, such as Joseph Dalton Hooker, botanist and explorer, who enthusiastically endorsed Darwin's theories. During this time Darwin was keenly aware that his theories needed to be as clear as possible and well supported by the evidence in biology. Also, his explanations for evolution were general, they applied to both plant and animal life, so that abundant examples were required for many species throughout biology.

Darwin used inductive methods of modern science. His theories were based on specific, detailed observations that supported his general conclusions about evolution. He went from specific, detailed, biological observations to a general hypothesis that became a theory. There was no guarantee that all observations would be supportive from the wide-ranging field of biology. He needed to consider all alternative and legitimate hypotheses and verify that his was the simplest (Occam's Razor) and was superior to all others in explaining all the data. His far-ranging work took almost twenty years.

In 1858, Darwin received a letter from Alfred Russel Wallace, a British explorer and naturalist. Wallace, also, observed thousands of species in various tropical environments and wrote an essay to Darwin with these conclusions: species evolve due to environmental pressures. Later, in his journal, Darwin wrote, "This essay contained exactly the same theory as mine."[11] Wallace's essay and an abstract of Darwin's own work were reviewed by other naturalists (peers) and Wallace's and Darwin's conclusions were published side by side in a journal for professional naturalists (Linnean Society of London, August 1858).

On November 22, 1859, *On the Origin of Species*, by Charles Darwin, was written, reviewed, and published. The original 1,250 printed copies immediately sold out. Darwin's theories were now in the public domain and they were widely spread throughout Europe and America. The most important controversy seemed to be that evolution was at odds with the biblical view of creation that we will examine in the next chapter.

11. Darwin, *Published Letters*, 82.

Darwin's *On the Origin of Species* was much debated mostly because many perceived the themes to be contradictory to what is in the book of Genesis, where God created through divine command all the life forms that exist on earth. Darwin's view of evolution and natural selection ran contrary to the Bible. In 1871, Darwin addressed the specific issue of how humans emerged through his theory of evolution when he published *The Descent of Man*.[12]

In this second book Darwin continued to focus on many issues raised in *On the Origin of Species* as they relate to humans. Darwin identified animal attributes and characteristics that were similar to human mental aptitudes. Darwin attempted to show the absence of a clear partition between animals and humans, especially in the areas of verbal communications, communal tasks and expectations, and basic morality. He also explained that many principal human qualities have naturally appeared due to mating struggles. The basic idea is that just like animals, humans emerged from the same natural processes that did not require a divine command God hypothesis.

Darwin has been portrayed as one of the most influential people in history and he was honored by being buried in Westminster Abbey.

CONCLUSIONS

As we have shown, Kepler's three laws of planetary motion can be understood only with the modern heliocentric theory and not the premodern geocentric model (see ch. 3). At the same time, the question remained about whether there is a more fundamental physical law that anticipates Kepler's three laws? It was Newton who answered this question when he formulated the universal law of gravitation through which he explained in precise mathematical terms how the gravitational forces that operate on earth are the same as those that keep the planets in orbit around the sun. Also, with his four rules for inductive reasoning, Newton gave helpful guidance for the practice of modern science.

We described how quantum physics (QP) deals with the very small objects and fields that are about the size of atoms. After being tested successfully, this theory has been used extensively in digital electronics. QP also helps us understand how and why the sun has been shining for the last 4.5 billion years. We now understand the processes that produced heavy atoms that are used as building blocks for all of life. However, unlike Classical Physics, a description of reality using QP is not very understandable using logic and deductions related to our everyday experiences.

12 Darwin, *Descent of Man*.

Another of the modern world's great scientists, Albert Einstein, advanced views that contradicted the accepted theories of his time. He assumed that the speed of light was constant regardless of how fast the observer was moving. His explanation that matter (mass) and energy are equivalent served as the theoretical basis for the terrible destructive power unleashed by atomic bombs. In 1915, Einstein predicted that the fabric of the four-dimensional space-time was warped due to the nearby presence of matter and/or energy. In 1919, his theory of general relativity was confirmed spectacularly by measurements of the bending of light near the sun.

The astronomer Edwin Hubble made meticulous observations and measurements of the cosmos using a state-of-the-art telescope that existed at that time. He convincingly showed that the universe extends beyond our own Milky Way galaxy and that the farther away a galaxy is from the earth, the faster it is moving. This led to his now widely accepted theory that the universe is expanding. Through additional observations his work led to the conclusion that the universe began 13.8 billion years ago when it was tiny and underwent a creation-like event called the big bang. Because of Hubble's scientific findings, we now recognize the enormous size of the expanding universe.

In addition to Newton, Einstein, and Hubble's discoveries of the how the physical cosmos came into being and operates, Charles Darwin offered a scientific explanation for the wide-ranging diversity of biological life forms that exist on earth. In his theory of natural selection, he detailed how species change according to competitive pressures they experience from their environments. He started by giving meticulous attention to the differences that exist among the diverse species of finches and other creatures that exist on the Galapagos Islands. After twenty years of continuing and careful research, he formulated a theory of evolution that applies to all earthly biological life.

Looking back over the past four chapters, starting with the work of Copernicus, modern inductive science has evolved a view of the universe that stands in stark contrast to that of our premodern ancestors. Through scientific experimentation, observation, analysis, peer review, and repeatability, we know now that the sun and not the earth stands at the center of our solar system. The stars are not pin pricks in an inverted bowl as they once seemed to be. Extraterrestrial objects do not move in perfect circles because they are in a pure and undefiled space.

The laws of motion and the law of gravity do not apply only to objects on earth but exist throughout the entire universe. Wavelike behaviors in the physical world are not completely separate from particle-like behaviors, and atoms do not undergo catastrophic collapse because their electrons fall into the atomic nuclei. Time and space are not separate, and the speed of light does not depend on the motion of the observer. The universe started 13.8 billion years ago, and the Milky Way is only one galaxy among the billions that exist throughout the expanding universe. All biological life that exists on earth, including modern humans, resulted from a process of natural sectional and evolution

It would not be an overstatement to say that the modern scientific view of the origin and development of the expanding universe as a whole, as well as how life evolved on earth, clashes with many of the traditional religious views that emerged during the premodern era. In the next chapter, we will examine the diverse responses that Christianity has made to this challenge.

5

Christianity and Modern Science

INTRODUCTION

In this chapter, we turn our attention to how Christianity has responded to the development of modern science. We start by revisiting chapter 1, where we listed the four big questions that we humans have been asking for centuries: (1) Where did the universe comes from and how does it operate? (2) Is there a spiritual power that transcends the universe? (3) If so, how is this spiritual power related to the universe? And (4) How is this power related to humanity? We begin our response to the first of these four big questions by summarizing chapters 2–4, where we discussed the differences between premodern and modern science. This will set the stage for the rest of the chapter.

In the premodern period, the prevailing perception of the universe was geocentric or earth-centered, where the sky appeared like an inverted bowl that the sun crossed by day and the stars by night. Its main support came from several sources, such as anyone's unaided-eye observations of the nighttime sky and a little imagination. Christianity reinforced this image because of its compatibility with the biblical account of creation. Also, astrologers and their followers made practical use of this image and the associated Ptolemaic theory by forecasting events in everyday lives of people, according to the heavenly movement of stars and planets. Because of its internal or theoretical coherence with the unaided eye and external or real-world applications, the premodern scientific view of the cosmos endured for more than twelve hundred years.

The year 1543 marks the beginning of modern science—when Copernicus challenged the earth-centered perception of the universe. After Copernicus, the combined work of many scientists led to the emergence of modern science, especially the contributions of Galileo, Kepler, Newton, Heisenberg, Einstein, Hubble, and Darwin. Here are some of the main elements of our current scientific understanding of the universe.

Our sun is the center of only our solar system, but it is not the center of the universe. Our sun is only one of billions of other suns that exist in our Milky Way galaxy that is only one galaxy among billions of other galaxies. The universe is very old, very large, and very complex. It began 13.8 billion years ago in the big bang. (See table of abbreviations at the beginning of this book—"bya" will be used for billion years ago.) It is 92 billion light years wide, and its expansion rate is accelerating. Our solar system with its planets began about 4.5 bya. Life on our planet earth began about 3.8 bya. We modern humans emerged about two hundred thousand years ago and have developed through a process of evolution based on natural selection and environmental adaptation.

None of this would have been possible without a shift toward inductive methods and the development of new technologies, starting with the telescope and the printing press. From the time of Galileo, a new way of thinking that includes observation, experimentation, measurement, predictability, and peer review became the standard for conducting modern scientific research for the past four centuries. As a result of this approach, our response to the first of the four big questions listed above about where the universe comes from and how it operates is that we have a modern scientific view of the universe that stands in stark contrast to its premodern predecessor.

With this background, we are ready to examine how Christianity has responded to modern science.

THE CHRISTIAN BIBLE

Three major issues lie at the heart of the relationship between Christianity and science. They are the role of the Bible, the doctrine of biblical inerrancy, and evolution. We start with the Bible since it is taken as ultimate authority by the faithful. Like all of the sacred scriptures that lie at the core of the world's many religions, in Christianity the Bible provides answers to the second, third, and fourth big questions about the existence of a transcendent spiritual power and its relationship to the physical universe and humanity. In answering the first big question, we have shown that the history of science can be divided into premodern and modern time periods and that the modern scientific understanding of when the universe began and how it operates is unexpected and vastly different from the premodern view.

The biblical description of how the world was created was written during the premodern era, and herein lies the challenging question: What is the relationship of the Christian faith to these two distinctly different views of creation? In terms of the four big questions, our answers to questions 2, 3, and 4 on the nature of God and God's relationship to the material universe and humanity depends on our response to question 1 regarding when and how the universe came into being and developed. It would not be an overstatement to say that starting with Galileo, during the past four hundred years, modern science, especially evolution, has challenged the Christian faith more than any other external force.

The reason is clear. If the modern scientific conclusion of how the universe came into being and developed is true, then the literal biblical creation story is false; and if the biblical account of creation is false, then might there be other areas of the Bible that are also false? For many Christians, this leads us onto a slippery slope. In answering these questions, Christianity has developed four different schools of thought on the relationship of Christianity to modern science. These four schools, which we discuss below, are Young Earth Creationism, Old Earth Creationism, Intelligent Design Creationism, and Evolutionary Creationism. Before we compare these four positions, however, we need to turn to the second of the key issues—biblical inerrancy—that has a direct bearing on how different groups of Christians understand the relationship of Christianity to science.

BIBLICAL INERRANCY

Biblical inerrancy means that the Bible is without mistake or fault in its entirety. The doctrine of inerrancy implies that nothing written in the Bible is contrary to fact. It is infallible.[1] Every word is literally true and authoritative in all matters related to the Christian faith. The current understanding of biblical inerrancy emerged during the past two hundred years in response

---

1. Because errors and faults were created during the process of copying and translating original manuscripts that no longer exist, the concept of inerrancy applies only to the original texts. While not everyone equates inerrancy with infallibility, many do. For this book, we assume that inerrancy and infallibility are synonymous. See https://defendinginerrancy.com.

to the rise of modern science, especially concerning the theory of evolution,[2] which we discuss in the next section of this chapter. For many Christians, the Darwinian interpretation of this theory that all forms of earthly life evolved over thousands of years, including *Homo sapiens*, threatened to undermine the biblical view that in six days God created by divine command all earthly species as they exist in their present form.

The doctrine of biblical inerrancy as it is currently understood and applied to science did not exist prior to the twentieth century. It is related directly to the rise of Christian Fundamentalism during the past one hundred years. In 1915, Milton and Lyman Stuart published a twelve-volume set of ninety essays called *The Fundamentals: A Testimony to the Truth*. The purpose of this publication was to specify the irrefutable truths without which they believed that the Christian faith would not be able to preserve its historical identity. In this publication, they established that the doctrine of inerrancy is one of the core truths.[3]

With the emergence of Fundamentalism starting in the early twentieth century, many Christians began rejecting and continue to reject several of the core concepts of modern science, especially when and how the universe started and developed. In essence, when the doctrine of biblical inerrancy is applied to creation, it leads to rejecting that the universe is 13.8 billion years old and that it evolved into its current form through progressive stages. It would not be an overstatement to say that in accepting a literalist—and therefore inerrant—view of the biblical account of creation, they are in essence denying the theory of evolution. Since the concept of evolution applies directly to so many of the discoveries of modern science, we need to discuss it in more detail in order to avoid confusion later when we compare Young Earth Creationism, Old Earth Creationism, Intelligent Design Creationism, and Evolutionary Creationism.

EVOLUTION

One of the key differences between premodern and modern science is evolution. During the premodern era, it was assumed that the universe as it was observed with the unaided eye in the form of an inverted bowl did not

---

2. A "theory" refers to a set of trusted scientific conclusions that are demonstrated with sufficient experimentation, resulting in a consensus in the scientific community. Further evidence in support of those conclusions is considered to be a waste of effort or funding because there are other pressing questions to address. For ideas in progress, the word "hypothesis" is used rather than "theory."

3 McFaul and Brunsting, *God Is Here to Stay*, 14–22.

evolve. God created the cosmic bowl at some point in the past, and to the unaided eye it appeared to remain unchanged. The world as observed was as it had always been and would always be. Modern science no longer accepts this view as we have shown in chapters 2–4. Instead, countless discoveries are best understood in this way: the world and life as it exists currently resulted from evolution. Thus, as modern science evolved from its premodern to modern views, the theory of evolution itself evolved starting with the big bang to the emergence of life on earth.

One of the best ways to understand evolution is to place it within the context of time and space. To repeat, the current scientific view of the age of the universe—13.8 billion years—resulted from the observations and simulations of Hubble and other modern-day astronomers and cosmologists about the size and expansion rate of the universe. For the finite human mind, a starting point occurring 13.8 billion years is virtually incomprehensible. Furthermore, since *Homo sapiens* did not exist when the cosmos began according to modern science, there is no way to verify this start date through direct human observation or experimentation. Determining when the universe began and how it evolved is the outcome of centuries of research, using many lines of inquiry, such as astronomy, quantum physics, and calculations based on known natural laws.

Once the universe began and started evolving, the next question relates to how it developed. In answering this question, we shift to space and then combine both time and space. Geological research has led to the conclusion that the entire biosphere of the earth has a thickness of at least six miles, extending above and below sea level. This translates into about a billion cubic miles.[4] Although this is an enormous volume of space, the theory of evolution applies to all the stages of change that have occurred and continue to occur over vast stretches of time (3.8 billion years) within this rather substantial volume of space (1 billion cubic miles).

In support of evolution, scientists have accumulated overwhelming evidences from many disciplines. Paleontologists and physical anthropologists have accumulated an abundance of fossil remains that show that the earth's life forms have evolved through many systematic and sequential changes. The observable record of the earth's crust clearly reveals a remarkable consistency from early fossils to more recent fossils. The strata that contain simple single-celled life forms predate strata that include multi-celled forms. These kinds of discoveries of evolutionary change on earth have recurred thousands of times.

4. The total volume of the biosphere can be estimated computing the earth's surface area: area = $4*pi*4000^2$ [set superscript 2] (radius of earth = 4000 miles) * 6 miles (thickness of the biosphere) = about 1 billion (not million) cubic miles.

There are other forms of evidence for evolution. Scientists have demonstrated that common physical structures exist among different species and that evolution explains these commonalities. For example, humans, mice, and bats have comparable skeletons that include bone-by-bone similarities. Upon comparing them, we notice that humans use arms to hug. Mice have limbs for scurrying about. Bat's flap their wings for flying. While these three species are vastly different, they all have similar structures.

In addition, evolution best explains the observable distribution of the 1,000,000 animal and 250,000 plant species that exist around the world. After countless generations of development, it is through the process of environmental adaptation that the earth's diverse animal and plant species have evolved into their current form. Despite these differences, the genetic code contained in their DNA is basically the same for all of them; and this, too, reinforces further the theory of evolution that scientists have tested, researched, and peer-reviewed extensively.

However, despite the extraordinary amount of research that strengthens the theory, evolution can never be proven with zero doubt. It is not a bulletproof and airtight theory that can identify all of the cause-and-effect relationships associated with the dynamics of change that drive the evolutionary process. For some critics, the theory is suspect because no one has ever witnessed precisely the step-by-step empirical stages that relate allegedly to changes that date back billions of years. At best, critics claim that the evidence used to support the theory of evolution is circumstantial.

At one level, this criticism is correct. Even through the sun-centered view of our solar system has replaced the premodern earth-centered view, Copernicus did not see directly the earth circling the sun. Nor did Kepler perceive the planets on their elliptical trajectories around the sun. Newton did not observe directly gravitational fields that hold the planets in their orbits around the sun and also causes apples to fall from trees. The same holds true for Einstein who did not witness directly the warpage of space-time or Hubble the galaxies beyond the Milky Way. Nonetheless, in terms of the cause-and-effect relationships that determine how the material universe operates, countless follow up experiments and mathematical analyses have confirmed repeatedly the reliability of the discoveries and conclusions that these scientists made.

This kind of recurrent confirmation of results does not exist in all areas of scientific research, especially when it focuses on how life emerged and developed on earth billions of years ago. Because the theory of evolution applies to about one billion cubic miles of space and a long period of time, it is not possible to conduct well controlled scientific experiments at all locations that lead to undisputed conclusions. Scientists do not have access to the exact environmental details for everything that happened over time and space since the universe started. While their work has yielded impressive results in deepening our understanding of the stages through which our universe evolved, our knowledge is far from comprehensive.

At the same time, modern science continues to expand and deepen our understanding of the evolutionary process through activities such as discovering new fossil remains or developing a new generation of powerful deep space telescopes. Also, simulations based on natural laws are improving. Results from interdisciplinary research are expanding. We are becoming better at acquiring and analyzing molecular and genetic data. While much more remains to be learned, thus far, the theory of evolution has been remarkably successful and predictive in explaining how life on our planet developed.

We are now ready to compare the four schools by which Christians define the relationship between Christianity and science and how they differ in their approaches to applying the authority of the Bible and the doctrine of biblical inerrancy with the theory of evolution. We will start with Young Earth Creationism and then follow with a discussion of Old Earth Creationism, Intelligent Design Creationism, and Evolutionary Creationism.

YOUNG EARTH CREATIONISM

Young Earth Creationism (YEC) assumes that the entire Bible is inerrant and must be interpreted literally. Therefore, it rejects the modern scientific theory of evolution. YEC holds to the premodern biblical view that God created the universe six thousand to ten thousand years ago. While this is an extraordinarily short time period, YEC contends that specific biblical passages support this time line, because the Bible is literally—and therefore factually—true. For example, according to the first three chapters in the book of Genesis, God created the universe in six days and then rested on the seventh day. How long is a day? Psalm 90 answers this question. In v. 4, the psalmist writes, "For a thousand years in your sight are like yesterday when it is past, or like a watch in the night." In the New Testament, 2 Peter 3:8 echoes this, saying, "With the Lord one day is like a thousand years, and a thousand years are like one day."

As we have shown in previous chapters, the modern scientific view of the age of the universe is vastly different from the YEC position.[5] To repeat, as a result of centuries of research by highly qualified scientists in fields such as physics, chemistry, biology, paleontology, anthropology, geology, and astronomy, modern science has shown that the universe began with the

5. Modern science gives a timeline that is 1.38 million times longer: 13.8 billion years / 10,000 years.

big bang 13.8 billion years ago (bya), that the earth formed about 4.5 bya, that life started on our planet about 3.8 bya, and modern humans (*Homo sapiens*) first appeared about two hundred thousand years ago. Thinking of these times as scaled to a year, the big bang occurs at 0 seconds on January 1, the earth forms eight months later on September 3, life originates on September 22, and *Homo sapiens* appear at just eight minutes before the end of the year at December 31, 23 hours, 52 minutes, and 59.5 seconds.

Despite several centuries of continuous observations and discoveries, the modern scientific perspective has not replaced the premodern biblical understanding in the minds of many present-day people. For example, the results of a 2017 Gallup survey about creationism showed that 38 percent of adults in the United States agreed with this statement: "God created humans in their present form at one time within the last 10,000 years."[6] Thus, despite the findings of modern science, how do we explain the persistent support for the premodern biblical view?

The answer to this question turns on the issue of authority. When YEC combines the total authority of the Bible with belief in inerrancy, then it follows that the biblical account of creation must be true. For modern science, the source of authority differs dramatically. Modern scientific authority is based on inductive research of the material universe to discover the laws of nature. Modern scientific research does not rely on a God hypothesis or reference to sacred texts. It involves observation, experimentation, measurement, predictability, and peer review, leading to the acceptance or rejection of a scientific theory about how natural laws operate. It is by embracing the authority of empirical methods of investigation that modern science has advanced from one discovery to another. In combination, these discoveries give us a very different picture of the cosmos that many times is nonintuitive.

Does this mean that Christianity is opposed to modern science? The answer is that it depends on how one interprets the Bible in relationship to the accepted theories of science. As we have indicated, there are four schools of thought on the relationship of Christianity and science: Young Earth Creationism, Old Earth Creationism, Intelligent Design Creationism, and Evolutionary Creationism. YEC is the most conservative. At its core, it is a binary or dualistic view. The Bible is assumed to be inerrant and infallible, which means that what the Bible says is literally true. All else, including the discoveries of modern science that contradict the biblical text, are therefore false.

In addition, the YEC perspective forbids tweaking or amending the text by redefining the language to make it seem like something it is not. For literalists, a day refers typically to a thousand years and not to any other time

6. *Wikipedia*, s.v. "Young Earth Creationism."

period. In addition, God created all of the plants, animals, humans, and physical objects in the order in which they are listed in the book of Genesis and not according to the progression of species that scientist working in the fields of paleontology, evolutionary biology, or physical anthropology have established. Thus, for those who accept YEC, we conclude that Christianity and modern science are completely incompatible.

However, this is not necessarily the case for the other three positions of Old Earth Creationism, Intelligent Design Creationism, and Evolutionary Creationism, which differ from YEC in how they interpret the Bible and apply it to the findings of modern science. We are now ready for Old Earth Creationism.

OLD EARTH CREATION

Both Old Earth Creationism (OEC)[7] and YEC hold to the doctrine of biblical inerrancy, but unlike YEC, OEC accepts the modern science understanding that the universe started with the big bang, developed through a process of evolution, and that the earth is 4.5 billion years old.[8] Consider a realistic example, such as the geological history of the Grand Canyon.[9] As the table indicates, the time line covers a period that is incompatible with the inerrancy assumptions that are made by YEC. As the table shows, according to event number 1, a flat-lying stratum accumulated on the Colorado Plateau 80 to 540 million years ago. Events 2–6 also record the time periods during which the Grand Canyon developed down to its present form.

Table 5. A Geological Summary of the Formation of the Grand Canyon

Event number	Millions of years ago	Summary of the event
1	80 to 540	The accumulation of the colorful, flat-lying strata on the Colorado Plateau
2	30 to 80	Laramide-age uplift and initial drainage to the northeast with a possible old Grand Canyon
3	16 to 30	Collapse of the Mogollon Highlands, development of the Chuska sand sea, and unknown drainage

7. Old Earth Creationism is also called Old Earth (Progressive) Creationism. In this book, we use only the designation Old Earth Creationism or OEC.

8. Stump, *Four Views on Creation*, 71–100.

9. Ranney, *Grand Canyon*, 141–66.

4	6 to 16	Lowering of the Basin and Range relative to the Colorado Plateau with interior or reversed drainage, or both
5	4 to 6	Complete integration of the Colorado River to the Gulf of California
6	4 to the present	Recent deepening of the Grand Canyon

All six of these events, along with their time boundaries (including uncertainties) that span millions of years, are totally inconsistent with the YEC conclusions. Contrary to YEC, however, the proponents of OEC could readily accept that the modern science of geology demonstrates that the origin of the Grand Canyon dates back to at least 80 million years. Because of its literalist inerrancy beliefs, as we have shown, YEC sets the creation of the entire universe at six thousand to ten thousand years ago. This means that it is necessary to deny completely the time line established by geological research. In order to explain these findings away, YEC followers would most likely assert that despite humanity's intellectual capacity to observe, measure, and understand the physical world, God has made it look like the formation of the Grand Canyon started 540 million years ago and created data to support this view.

In addition, OEC departs from YEC's plain-words, literalist reading of the Bible for the following reason. Biblical scholars have noted that the meaning of biblical words changed over time depending on the different historical contexts in which they were used. Language evolved and was expressed flexibly. As we have shown above, for YEC, both Ps 90:4 and 2 Pet 3:8 indicate that for God, one day is like a thousand years. YEC and OEC would differ in their interpretations of these two passages. For YEC, the phrase "like a thousand years" means literally that one day equals a thousand earth-calendar years. Following the biblical language patterns, the OEC position allows for more flexibility. "Like a thousand years" could be interpreted to mean that for God, a single day is like a very long time.

While this might sound like the OEC advocates really do not believe in inerrancy, this is not necessarily the case. Here is why. There are other Bible references that illustrate this point. The biblical description of the length of time that the ancient Egyptians enslaved the Israelites was four hundred years (Gen 15:13 and Acts 7:6). When they escaped under Moses' leadership, they wandered in the desert for forty years (Deut 2:7; 8:2; 29:5; Num 32:13). After John baptized Jesus in the Jordan River, Jesus spent forty days and nights in the wilderness (Matt 4:1–2).

The references to numbers such as four hundred and forty parallels the biblical use of the image that one day equals a thousand years. For YEC, references to biblical events and the language used to describe them should be interpreted literally. For OEC, only the events themselves are to be viewed literally as having occurred, but the language that describes them can be understood metaphorically or symbolically. Thus, for OEC, God literally created and developed the universe over a very long time; the Jews were held as Egyptian slaves for a very long time; after escaping, the Jews wandered in the desert for a very long time; and Jesus retreated into the desert for a very long time.

Given the absence of modern calendars and time-keeping methods, for OEC biblical allusions to time periods such as four hundred years, forty years, and forty days could be interpreted easily to mean "for a long time." Even though both OEC and YEC hold to the doctrine of inerrancy, OEC does not accept the YEC perspective that the earth is only six thousand to ten thousand years. Thus, unlike the disagreements that exist between YEC and modern science, there is no conflict between OEC and modern science on the age of the material universe (13.8 bya), the earth (4.5 bya), and life on earth (3.85 bya).

However, this does not mean that OEC agrees with all other discoveries of modern science. There is one area in particular where incompatibility between OEC and modern science stands in high relief: evolution. From the current perspective of Darwinian theory, the emergence and development of life on earth is the result of an uninterrupted genetic continuum. This is in response to each species becoming adapted for survival in a specific environment in competition with other species. OEC does not accept this view because it maintains that all forms of earthly life appeared suddenly through a divine command sequence of creation events. In other words, while OEC rejects as literally true the YEC position that the universe is six thousand to ten thousand years old, it accepts as literally true that God created life on earth through divine command.

The key to understanding OEC's rejection of the theory of evolution is based on the existence of creation gaps that appear in the evolutionary record of how one species emerged out of another. Creation gaps exist where there is an absence of empirical data showing a direct and verifiable cause-and-effect relationship between earlier species and later ones. Added to this is the OEC position that there is no definitive evidence that there was an original common ancestor from which any given species evolved. Here is an example that illustrates this point. In 1974 in Ethiopia, paleontologist Donald Johnson discovered 40 percent of the skeletal remains of a female, who has come to be known affectionately as Lucy. After comparing the bone structure of Lucy with other unearthed skeletal evidence, he concluded that she lived 3.2 million years ago and was the original mother of many hominoid species that no longer exist and from which *Homo sapiens* evolved eventually.

Because of the enormous time gap and unproven causal connections between extinct hominoid as well as their relationship to each other and humanity, OEC holds that this kind of reasoning has little merit. Unlike the earlier example of the Grand Canyon, where the geological time sequences, erosion patterns, and drainage levels can be identified with a high degree of scientific certainty, this is not the case with Lucy and subsequent hominoid species that allegedly evolved into *Homo sapiens*. Asserting that the fossil record supports the claim that there are causal connections from Lucy to humanity is speculative at best. Due to the existence of so many creation gaps that cannot be connected definitively and scientifically, and probably never will be, OEC leads to an acceptance of the biblical view that God created all earthly species in their present form through divine command after the earth was formed 4.5 bya.

The vast majority of scientists do not accept the divine command theory of species creation because the OEC position does not lead to advancing our understanding of how life started and developed from an earliest common ancestor. Accepting as literally true the Bible's creation narrative would mean that the different species on the Galapagos Islands that Darwin studied came into being through distinct, unconnected, and unnatural creation events. Nowhere in the scientific literature or research that is based on shared observations, experiments, empirical tests, predictability, and peer review is there any evidence of such brief and limited divine command occurrences in the origin of finches or any other species.

Thus, what we find when comparing OEC with YEC is that YEC applies the doctrine of inerrancy comprehensively to the Bible as a whole, whereas OEC does so only selectively. While OEC accepts the modern scientific understanding of when the universe and earth started 4.5 bya, it rejects the theory of evolution of life on earth. Instead, OEC maintains that God created all earthly species as they exist currently through divine command. Another way to say this is that YEC applies a strictly literal interpretation of the Bible in full, whereas OEC does so only in part. How does the third school called Intelligent Design Creationism differ from YEC and OEC in its understanding of the relationship between Christianity and modern science?

INTELLIGENT DESIGN CREATIONISM

Like YEC and OEC, Intelligent Design Creationism (IDC) accepts the authority of the Bible and doctrine of inerrancy but applies them differently to the theory of evolution. IDC is more akin to OEC than YEC, because it is only a partial application of the doctrine of inerrancy to the findings of modern science. As we show below, IDC rejects the theory of evolution, but it does so for reasons that differ from both YEC and OEC. Before we discuss where this difference exists, there is one additional issue that motivates the followers of all three position. At the heart of their rejection of evolution is the belief that this theory leads to atheism.

It is a slippery slope argument that can be summarized as follows: If we start with evolution today, then we will end up with godless atheism tomorrow. Little by little, evolution whittles away at belief in God. Why should this be so? The answer to this question is that the constant uncovering of more and more verifiable discoveries has resulted in evolution emerging as a key theory in biology and that this will result ultimately in a godless, random, and unguided explanation of how any and all life forms began. In turn, this will eliminate the need for the biblical belief that God created the universe, the earth, and all of the plant and animal species that exist on earth through divine command. Therefore, evolution must be challenged by believers in God; and in addition to YEC and OEC, IDC is one of the best ways to do this.

As we have shown, YEC adheres to a literalist interpretation of the whole Bible and rejects the findings of modern science on the beginning of the universe, the earth, life on earth, and modern humans. OEC holds to this general modern scientific timeline but disagrees with the process by which each of the stages appeared. Rather than accepting a naturalistic explanation based on evolution, OEC opts for the biblical alternative of divine command based on the concept of creation gaps. IDC shares this divine command interpretation with OEC but with a different emphasis. It adds to the creation gaps perspective the notion of complexity. In each of these three cases, the path to atheism is avoided.

For the IDC perspective, the theory of evolution is flawed because it cannot explain many of the intricacies of nature. IDC proponents maintain that some species and their functions are so irreducibly complex that they could not possibly have come into existence through random and unguided genetic mutations that were simpler in earlier species. These species could exist only through the work of an intelligent designer who brought them into being through divine command. Thus, it is through the combination of the concepts of creation gaps and irreducible complexities that both OEC and IDC continue to reject the theory of evolution.

Instead, God created the universe and all subsequent stages of complex development *ex nihilo*, that is, out of nothing.

Thus, despite the abundance of discoveries that modern science has made in demonstrating the validity of evolution from the start, many conservative Christians began challenging proponents of evolution to explain (1) the precise causal connections through which later species developed from earlier ones and (2) the countless complexities by which nature operates. This is the problem Darwin encountered when he first proposed evolution as the process through which new varieties of plants and animals emerged. During his time, there was no known natural cause-and-effect explanation for how genetic information was transferred from generation to generation or how genetic mutations occurred. As a result, his critics maintained that the theory of evolution is untenable.

The strength of IDC is that it recognizes correctly that the current theory of evolution does not have rigorous, data-based explanations that show all the causal connections between later and earlier species. However, these gaps are closing. At the same time, IDC is vulnerable to criticism at two levels. The first is scientific in that IDC does not provide testable predictions for observations. Predictability is a hallmark for successful scientific progress. The second is that it does not provide any kind of materialistic mechanism for the assumed divine interventions.[10] In combination, these two criticisms expose the weakness of IDC, and by implication YEC and OEC, as well. Successful discoveries are the result of applying inductive methods to observations. This includes the development of theories that lead to falsifiable hypotheses about how cause-and-effect relationships operate in the material universe. This process involves repeated observations and experiments that are peer reviewed in order to determine whether they are true or false. The end point of this process is predictability.

For example, the big bang theory predicted cosmic microwave background (cmb) radiation with surprising accuracy. The prediction was that the observed radiation (microwaves) would be of equal intensity from all directions and the radiation would have a specific cold temperature. These predictions were unknown to the investigators who through repeated experimentation ultimately found the cmb.[11] This prediction and subsequent experimentally-based findings strongly supported the now widely accepted big bang theory, and the competing steady state theory was rejected. In this case, considerations of the big bang theory implied subsequent

10. Collins, *Language of God*, 187–88.

11. See chapter 11, "Christianity and Four Big Questions," subsection 5, "What is the fate of the universe?"

measurements (predictions) that might confirm or falsify it. Furthermore, as we have shown in chapters 3 and 4, it was through repeated and peer reviewed observations and experiments that the sun-centered understanding of our solar system replaced the centuries old earth-centered theory.

Unlike those experiments, critics of the theory of evolution claim correctly that it is not possible for modern science to duplicate through any kind of direct experimentation the process of evolution that is purported to have occurred over billions of years. As stated above, during the mid-nineteenth century, Darwin confronted this issue when he first advanced his theory of evolution, and there did not exist at that time a way to explain how later species emerged from earlier ones or why diversity existed within any given species, such as different types of finches. Knowledge about the genetic structures of life that is encoded in DNA did not exist at that time.

In 1953, this changed when James Watson and Francis Crick announced the discovery of the structure of DNA which is the physical basis for genetic information for all forms of life. This gave additional scientific support for the theory of evolution.[12] New species emerge when genes are transferred from generation to generation. This transfer process also applies to reproductive mutations that account for variety within a species.

Does the discovery of DNA prove the theory of evolution? The answer is no. However, in combination with earlier scientific findings, discovering that DNA exists and how it operates in nature reinforces the theory of evolution and enhances confidence that it is correct. The reason for this is that it provides a scientific basis for explaining how later species evolved from earlier ones through the transfer of genetic information and how mutations lead to variety. It is one more piece of evidence, unknown to Darwin, that leads to closing the creation gaps between generations. Simply stated, the discovery of DNA has given additional support for evolution of life on earth, which stands in stark contrast to the divine command position that cannot explain how God's intervention led to the development of life at the atomic, molecular, and genetic levels.

Furthermore, scientific studies of complex natural areas that IDC claims that only God could have created by divine command have yielded to scientific cause-and-effect explanations, such as the operations of the human eye. As science continues to advance in its discovery of the natural laws that govern the material universe, the spaces where God is assumed to operate, such as in creation gaps and the intricacies of nature, begin eventually to shrink and disappear.

---

12. Schoub, *Seeing God through Science*, 36–39.

Does the discovery of DNA and how it operates in nature, along with a huge and growing number of other scientific findings that support the theory of evolution, take us down the slippery slope that ends with atheism? The answer is that it depends on how we interpret the Bible and its premodern view of creation. For Christians who accept the doctrine of inerrancy, there are two possibilities. For Young Earth Creationism, the Genesis account of creation must be viewed as literally true based on a plain-words interpretation that one day equals a thousand years. For Old Earth Creationism and Intelligent Design Creationism, the biblical language can be interpreted metaphorically and thus accommodated to the modern science creation timeline that goes back 13.8 billion years.

At the same time, for the followers of these positions, a deeper concern surrounding the theory of evolution is twofold. The first is that evolution replaces the doctrine of divine command. The second is closely related to the first. Rejecting the biblical emphasis on divine command as the basis for explaining the emergence of new species apparently leads to discarding the need for a God hypothesis to describe how life on earth started and developed. Stated differently, the theory of evolution results in a completely naturalistic understanding of the universe; and this is the essence of atheism.

IDC counters a naturalistic explanation for how the cosmos began and developed by contending that the universe is so finely tuned that it infers, but does not prove, the existence of an intelligent designer who brought it into being and guided its evolution. Stephen C. Meyer, opinion leader for IDC, acknowledges that "the design we observe in nature is real, not just apparent." Furthermore, "Though intelligent design is not *based upon* religious belief, it does affirm a key tenet of a biblical worldview—namely, that life and universe are the products of a designing intelligence—an intelligence that I, and other Christians, would attribute to the God of the Bible."[13] Thus, IDC complements both YEC and OEC by emphasizing that God's footprint is evident in our highly structured universe along with its irreducible complexities for which no currently accepted scientific explanations exist.

Next, we turn to the fourth and final position of interpretation on the relationship between Christianity and science: Evolutionary Creationism.

---

13. Stump, *Four Views on Creation*, 207.

EVOLUTIONARY CREATIONISM

For Evolutionary Creationism (EC), there is no conflict between Christianity and modern science. The proponents of EC reject Young Earth Creationism that views every word of the whole Bible as literally true. EC also rejects a metaphorical interpretation of the biblical language in order to preserve the doctrine of divine command creation for the emergence of new species or variety within a species. Simply stated, EC accepts completely the modern scientific view of the big bang and evolution.

The EC position is summarized by Deborah B. Haarsma, who is president of BioLogos, an Evangelical Christian advocacy organization. In her words, "Evolution is not a worldview in opposition to God but a natural mechanism by which God providentially achieves his purposes."[14] It is precisely through the process of evolution that God governs the universe. "Evolutionary creation is the view that God created the universe, earth, and life over billions of years, and that the gradual process of evolution was crafted and governed by God to create the diversity of life on earth."[15]

Because of the central importance of the Bible for the Christian faith, the proponents of EC recognize that its main challenge is to reconcile the Bible with evolution without sliding down the slippery slope that leads to atheism and rejection of the entire Bible because it was written during the premodern scientific era. In so doing, EC does not reject modern science as YEC does or seek a partial reconciliation as OEC and IDC do. Instead, EC accepts in full that the modern scientific view of how the universe started and developed is accurate.

The EC position is that the Holy Spirit inspired the biblical writers but that they lived in diverse cultures, spoke with different languages, and used many dissimilar literary styles and genres. It took more than twelve hundred years to compile the thirty-nine books of the Old Testament and twenty-seven books of the New Testament. All of the biblical writers lived during the premodern era that was based on a static, inverted bowl image of the heavens. They viewed the earth as flat and at the center of the universe, which we described in chapter 2. For EC, the modern scientific view of the universe has replaced this view without eliminating belief in God. If we assume the Bible had been written with a modern scientific viewpoint, premodern readers would not have understood or accepted the text.

14. Stump, *Four Views on Creation*, 125. After his conversion to Christianity, Francis S. Collins, a top world biologist identified with the Human Genome Project and the National Institute of Health, founded BioLogos in 2007. Biologos is a Christian advocacy group that attempts to reconcile evolution with the idea of God.

15. Stump, *Four Views on Creation*, 125 (italics in original).

As a result, EC acknowledges that the Bible is not a modern scientific textbook and was never intended to be. "Most evolutionary creationists do not see the Bible as making scientific predictions or referring to science unknown to the ancient readers."[16] What is important for EC is not the literalism or inerrancy of biblical words about creation, however translated or tweaked, but the deeper theological meaning underneath them. From the perspective of the Bible, there is only one God who is the Creator of all that exists. Modern science does not eliminate belief in God but instead reinforces it. "Our view of nature as God's creation is what gives us the underlying motivation to pursue science and the ultimate praise for its discoveries . . . For the Christian, a scientific explanation glorifies God by revealing his handiwork."[17] If the biblical witness exists to glorify God's creative handiwork, then for EC advocates, modern science's discoveries of the big bang and evolution, both physical and biological, fulfill this theological purpose more faithfully and truthfully than the Bible's premodern story of creation.

As we have shown, YEC, OEC, and IDC accept the doctrine of inerrancy, even though they apply it differently to their interpretation of the Bible. Does this mean that by accepting modern science's support for evolution from the time of the big bang forward that EC rejects biblical inerrancy? The answer is no. EC retains the doctrine of biblical inerrancy but does not apply it to the details of Genesis story of creation in any form. However, in other biblical areas, EC retains the doctrine of inerrancy. "Evolutionary creationism (EC) joins with YEC [and OEC and IDC] in believing that the God of the Bible is the creator of the entire universe and that Jesus Christ is the only path to salvation."[18]

Thus, the proponents of all four positions (YEC, OEC, IDC, and EC) share this core Christian conviction about salvation, also called soteriology, regardless of how they differ in applying the doctrine of inerrancy to the Bible's premodern view of creation. On the issue that Christ is the only means to salvation, they stand united. Before turning to the next chapter that compares the four schools according to the standards of modern science, we conclude with a summary of how they differ in their understanding of the relationship between Christianity and science.

16. Stump, *Four Views on Creation*, 131.

17. Stump, *Four Views on Creation*, 130.

18. Stump, *Four Views on Creation*, 55; bracketed note added.

CONCLUSIONS

The following table summarizes the four different position that we have compared in this chapter.

Table 6. Summary of Four Views of Christianity and Science

Christian position	Acceptance of modern theory of physical evolution and timeline?	Acceptance of modern theory of biological evolution and timeline?	Acceptance that Jesus Christ is the only path to salvation?
Young Earth Creationism (YEC)	No	No	Yes
Old Earth Creationism (OEC)	Yes	No. Based on creation gaps in the biological record	Yes
Intelligent Design Creationism (IDC)	Yes, plus finely tuned universe implies a creator God	No. Based on nature's irreducible complexities	Yes
Evolutionary Creationism (EC)	Yes	Yes	Yes

As we have shown, the proponents of all four positions combine the Bible, the doctrine of inerrancy, and the theory of evolution in different ways. The left side column 1 lists the four positions for defining the relationship between Christianity and science along with their abbreviations. Columns 2 and 3 summarize how the four positions differ in accepting the modern scientific timeline for the start of the universe and for physical and biological evolution.

YEC rejects both columns 2 and 3 that apply to the modern science timeline and the belief in physical and biological evolution. The other three positions of OEC, IDC, and EC accept the timeline along with physical evolution in column 2. However, as column 3 reveals, only EC accepts biological evolution. OEC and IDC reject biological evolution, because they hold that God created all the species that exist on earth through divine command. They cite the existence of creation gaps and the complexities of nature to support their position. IDC adds that the finely tuned nature of the universe implies the existence of a creator God. Even though the proponents of the four positions disagree on how to combine the doctrine of biblical inerrancy to the Genesis story of creation, column 4 indicates they are in complete agreement that the Bible is inerrant on the crucial claim that Christ is the only path to salvation.

We will return to the doctrine of inerrancy in the second part of this book (chs. 7–10), which deals with Christianity and religious pluralism. However, before doing so, we turn next to chapter 6, where we will examine how each of the four positions compares with modern science.

6

Four Christian Positions
and Modern Science

In this chapter, we make direct side-by-side comparisons of the four positions we discussed in the previous chapter with thirty-two key features associated with the methods of modern science. To reiterate, these four are Young Earth Creationism (YEC), Old Earth Creationism (OEC), Intelligent Design Creationism (IDC), and Evolutionary Creationism (EC). Such a comparison will help identify areas of overlap between materiality (level 1) and spirituality (level 2) that we included in Table 2, chapter 1. Our reference for modern science (cell 3 in Table 2) is the National Academy of Sciences; and our summary of the four Christian positions (cell 4 in Table 2) is taken from *Four Views on Creation, Evolution, and Intelligent Design.*[1]

ASPECTS OF MODERN SCIENCE

Before we do a detailed comparison of the four Christian positions and modern science, it is helpful to identify some of the most important aspects of modern science that have emerged since 1543 CE. The first is falsifiability, which refers to the possibility that a scientific theory or hypothesis could be contradicted by verifiable evidence. For example, the idea of the sun, moon, and stars encircling the earth because the earth and humankind were the center of the universe was falsified by the Copernican heliocentric universe

---

1. Stump, *Four Views on Creation.*

and the subsequent telescopic observations by Galileo. We will show how falsifiability relates to each of the four positions.

Scientific conclusions and demonstrations are typically based on experiments and observations. A scientific theory refers to a set of trusted scientific conclusions that are demonstrated with sufficient experimentation, resulting in a consensus in the scientific community. Further evidence in support of those conclusions is considered unnecessary because there are other important questions to investigate. For ideas that are in the process of investigation, the word "hypothesis" is used rather than "theory." In modern science, evolution is defined as a theory because it is well understood, and there is much physical and biological evidence to support it.

Modern science also tends to be self-correcting. For example, as we discussed in chapter 4, the taken-for-granted scientific idea (also called logical or intuitive) that the measured speed of light depends on the speed of the observer relative to the light source was found to be false. In a scientifically self-correcting moment, Einstein assumed that the measured speed of light was independent of the speed of the observer in his theory of special relativity. This nonintuitive correction has been experimentally confirmed many times.

Peer review is essential for self-correction in modern science. When a scientist has evidence that points to a new hypothesis. Those evidences and explanations are documented typically and sent to a trustworthy and peer-reviewed scientific journal for open publication. As a result, others around the world might learn and benefit by this new information. Also, a scientist might have overlooked some experimental details that other scientists could identify during the peer review process. For example, in 1953 James Watson and Francis Crick discovered the double-helix structure of DNA and published their findings in *Nature*.[2] This historic paper was peer reviewed before publication and ultimately led to a Nobel Prize in 1962.

Using these basic criteria, we will examine how each of these four positions compares with the findings of modern science. In the first chapter, we identified four big questions about how the material universe operates and the nature of a spiritual power that might transcend it or be immanent within it. In this chapter, our goal is to compare the four positions of YEC, OEC, IDC, and EC with these four big questions in order to determine which position is most consistent with modern science. In order to achieve this goal, we turned to the National Academy of Sciences and its publications to develop a list of twenty elements (twelve more were added for a total of thirty-two) that modern scientists follow to conduct their research.

2. Watson and Crick, "A Structure for Deoxyribose Nucleic Acids," 737–38.

Based on this list, we have developed a comparison table and an easy-to-understand bar chart, Figure 24, that summarizes our comparison.

NATIONAL ACADEMY OF SCIENCES

The National Academy of Sciences (NAS) consists of members, each of whom is a distinguished scientist and who makes continuing important contributions in their original scientific research.[3] One of the highest honors any scientist can have is be elected to the NAS. About 190 members of the NAS (out of a total of about 2,350 members) have won a Nobel Prize, viewed by many as the highest level of scientific recognition in the world. It can be safely concluded that the NAS is the most informed and trustworthy resource in the United States for their conclusions from modern science. Here is the mission statement of the NAS:[4]

> The National Academy of Sciences (NAS) is a private, non-profit society of distinguished scholars. Established by an Act of Congress, signed by President Abraham Lincoln in 1863, the NAS is charged with providing independent, objective advice to the nation on matters related to science and technology. Scientists are elected by their peers to membership in the NAS for outstanding contributions to research. The NAS is committed to furthering science in America, and its members are active contributors to the international scientific community. Approximately 500 current and deceased members of the NAS have won Nobel Prizes, and the Proceedings of the National Academy of Sciences, founded in 1914, is today one of the premier international journals publishing the results of original research.

NAS provides the most authoritative summary of what is modern science, especially as it relates to evolution. We assume that the NAS is the most qualified authority on the nature and methods of modern science.[5] Thus, we use the NAS guidelines for comparing and evaluating the four positions. The National Academy of Sciences list of standards includes twenty factors related to the correct conduct of modern science. We have added twelve more factors to this list of twenty for a total of thirty-two criteria that serve as the basis for the following comparison.

3. See http://www.nasonline.org.

4. See http://www.nasonline.org/about-nas/mission.

5. One of the most applicable NAS references, used here, is *Science and Creationism: A View from the National Academy of Sciences* (2nd ed., National Academy of Sciences, 1999).

CHRISTIANITY, MODERN SCIENCE, AND THE FOUR REACTIONS

In Table 7 below, we have specified thirty-two modern science characteristics (in thirty-two rows) that we use to compare YEC, OEC, IDC, and EC. Each of these separate positions appears in one of the four columns. For the entire table, the total number of cells is 128 (32 x 4 = 128). Each cell has been scored with a 0, 1, 2, or 3. As Figure 23 shows, a score of 0 indicates there is complete (or nearly complete) disagreement between a given position (1 of 4) and the specific characteristic (1 of 32 rows). A score of 1 indicates that there is more disagreement than agreement between that characteristic and the modern science characteristic but not complete disagreement. A score of 2 indicates that there is more agreement than disagreement for the position and characteristic. A score of 3 indicates there is complete (or nearly complete) agreement between a given position (1 of 4) and the specific characteristic (1 of 32 rows).

0 – 3 scale, used to score each of the 4 positions

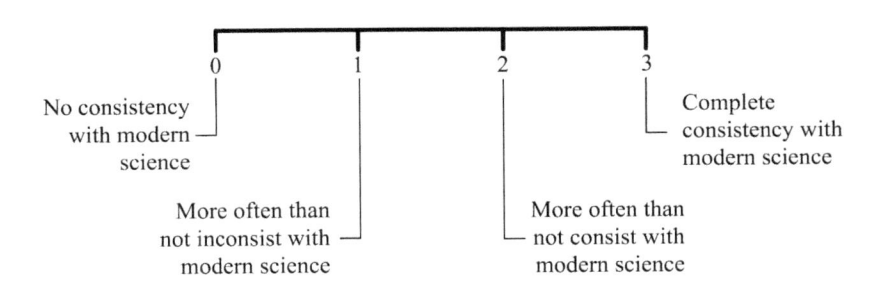

Figure 23. Scale to score each of the four positions with each of the thirty-two characteristics of modern science.

In the following Table 7, we placed a numerical score in each of the 128 cells after reviewing very carefully how each of the thirty-two characteristics applies to each of the four positions. We made our scoring as fair and unbiased as possible. Table 8 explains why we placed a given score ranging from 0 to 3 for each of the four positions according to the characteristics that appear in the thirty-two rows.

Table 7. Scoring of each of the four positions (columns) with each of the 32 characteristics of modern science (32 rows). Justification for the scoring is given in Table 8.

	Good modern science does this:	YEC	OEC	IDC	EC
1	Explains natural phenomena with natural based causes.	0	1	2	3
2	Fosters technical innovation.	0	1	2	3
3	Relationships and explanations between natural phenomena are sought.	0	1	1	3
4	Explanations help understand various kinds of phenomena.	0	1	2	3
5	Explanations can be inferred from careful observations and experiments that can be substantiated by other scientists.	0	1	2	3
6	Explanations must be supported by enough evidence.	0	1	2	3
7	Explanations are empirical. *	0	1	2	3
8	Explanations lead to testable deductions.	0	1	2	3
9	Frequently one natural phenomenon can be related to another.	0	1	2	3
10	Natural phenomena have natural causes.	0	1	2	3
11	Actively seeks applicable evidences.	0	1	1	3
12	Evidences can be subjected to meaningful tests.	0	1	1	3
13	Detailed explanations are developed, carefully documented, elaborated, and shared.	0	1	2	3
14	Increasingly better explanations of natural phenomena are encouraged.	0	1	1	3
15	Thorough testing of explanations is encouraged.	0	1	1	3

16	Some accepted hypotheses are found to be weak or incorrect when subjected to additional testing and then rejected.	0	0	0	3
17	Many explanations are so thoroughly tested and confirmed that they are held with great confidence.	0	0	0	3
18	Unifying explanations apply to multiple evidences.	0	1	2	3
19	Recently obtained evidences are allowed to support, alter, or reject current explanations.	0	0	0	3
20	Some explanations are not directly observed but are inferred from available observations.	0	1	1	3
21	Explanations are peer reviewed before they are presented and/or published in the scientific literature.	0	1	1	3
22	Competing natural explanations are openly discussed and evaluated by knowledgeable and unbiased reviewers.	0	0	0	3
23	Confirming or falsifying observations are encouraged.	0	0	0	3
24	Explanations can be falsified.	0	0	0	0
25	Falsified explanations are rejected in favor of ones that agree with all applicable observations.	0	1	1	2
26	Many explanations are quantifiable.	0	1	2	3
27	Comprehensive observations and testing of explanations are encouraged.	0	1	2	3
28	Some explanations are rejected and not published or presented.	0	1	2	3
29	Successful explanations that predict recent observations are strengthened.	0	0	0	3
30	Predictions from accepted explanations are compared to the data.	0	1	2	3

31	Evidences are meticulously observed and documented without bias.	0	1	2	3
32	Predictions from competing explanations are compared with the data.	0	0	0	3

* "Empirical" here refers to understandings that are based on observations or experiences.

Table 8. Justifications for the scoring in Table 7.

Row number	Summary of justification.
1	YEC is 0 because all natural phenomena are explained only with inerrant literal biblical interpretations. OEC is 1 because physical evolution is accepted but not biological evolution. IDC is 2, because physical evolution is accepted some, but not all, of biological evolution is accepted. EC is 3, because all of physical and all of biological evolution is accepted.
2	Same justification as for row 1 above.
3	YEC is 0 because there is no room for anything but inerrant biblical literalism. OEC is 1 and IDC is 1 because only physical evolution apples. EC is 3 because both physical evolution and biological evolution apply and everything is open for this.
4	YEC is 0 because there is no room for anything but inerrant biblical literalism. OEC is 1 because only physical, but not biological phenomena are open for explanation, leaving out all of health care. IDC is 2 because physical, but not all biological phenomena are open for explanation. EC is 3 because both physical evolution and biological evolution apply and everything is open for this.
5	Same justification as for row 4 above.
6	YEC is 0 because there is no room for anything but inerrant biblical literalism. OEC is 1, IDC is 2, and EC is 3 for the same reasons as in #4.
7	Same justification as for row 6 above.
8	Same justification as for row 4 above.

9	YEC is 0 because there is no room for anything but inerrant biblical literalism. OEC is 1 because only phenomena in the physical space applies. IDC is 2 because much, but not all, phenomena in the biological space applies, leaving out healthcare. EC is 3 because both physical evolution and biological evolution apply and everything is open for this.
10	Same justification as for row 9 above.
11	YEC is 0 because there is no room for anything but inerrant biblical literalism. OEC is 1 and IDC is 1 because large spaces in biology are off limits. EC is 3 because both physical evolution and biological evolution apply and everything is open for this.
12	Same justification as for row 11 above.
13	Same justification as for row 9 above.
14	Same justification as for row 11 above.
15	Same justification as for row 11 above.
16	YEC is 0 because there is no room for anything but inerrant biblical literalism. OEC is 0 and IDC is 0 because biblically based hypotheses are not allowed to be weak. EC is 3 because all of both physical evolution and biological evolution hypotheses apply.
17	Same justification as for row 16 above.
18	Same justification as for row 11 above.
19	YEC is 0 because there is no room for anything but inerrant biblical literalism. OEC is 0 and IDC is 0 because no recently obtained evidences in the inerrant biblical literal spaces are allowed. This is a "God of the gaps" type argument. EC is 3 because all of both physical evolution and biological evolution hypotheses apply.
20	YEC is 0 because there is no room for anything but inerrant biblical literalism. OEC and IDC are 1 because gaps in biological evolution are interpreted as places where God must have acted. EC is 3 because all of both physical evolution and biological evolution hypotheses apply.
21	YEC is 0 because there is no room for anything but inerrant biblical literalism. OEC and IDC is 1 because peer review applies only for those spaces that are not associated with a biblical literal interpretation. EC is 3 because all of both physical evolution and biological evolution hypotheses apply.

22 YEC is 0 because there is no room for anything but inerrant biblical literalism. OEC is 0 and IDC is 0 because biblically based hypotheses are not allowed to be altered in any way. EC is 3 because all of both physical evolution and biological evolution hypotheses apply.

23 YEC is 0 because there is no room for anything but inerrant biblical literalism. OEC is 0 and IDC is 0 because biblically based hypotheses are not allowed to be falsified. EC is 3 because all of both physical evolution and biological evolution hypotheses apply.

24 YEC is 0 because there is no room for anything but inerrant biblical literalism. OEC is 0 and IDC is 0 because biblically based hypotheses are not allowed to be considered for falsification. EC is 0 because Jesus is the only path to salvation cannot be falsified. (See Table 6 in chapter 5.)

25 YEC is 0 because there is no room for anything but inerrant biblical literalism. OEC is 1 and IDC is 1 because biblically based explanations are not allowed to be challenged but other spaces are acceptable. EC is 2 because there are larger, but not total spaces, apply.

26 YEC is 0 because there is no room for anything but inerrant biblical literalism. OEC is 1 and IDC is 2 because explanations for large spaces in biology are off limits and considered for quantification. EC is 3 because all of both physical evolution and biological evolution quantification apply.

27 YEC is 0 because there is no room for anything but inerrant biblical literalism. OEC is 1 and IDC is 2 because there are large spaces that are off limits for this. EC is 3 because all of both physical evolution and biological evolution quantification apply.

28 YEC is 0 because there is no room for anything but inerrant biblical literalism. OEC is 1 and IDC is 2 because if the disallowed spaces are not considered, then it is acceptable to reject those explanations. EC is 3 because all of both physical evolution and biological evolution quantification apply.

29 YEC is 0 because there is no room for anything but inerrant biblical literalism. OEC is 0 and IDC is 0 because recent observations in disallowed space are not allowed to be considered. EC is 3 because all of both physical evolution and biological evolution quantification apply.

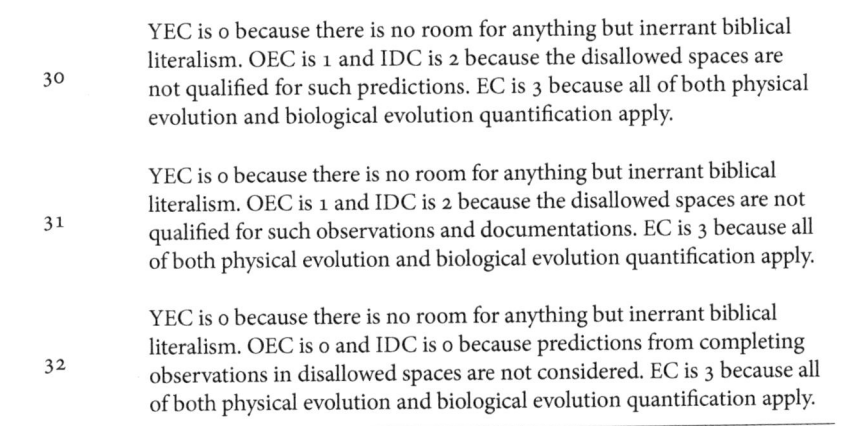

30	YEC is 0 because there is no room for anything but inerrant biblical literalism. OEC is 1 and IDC is 2 because the disallowed spaces are not qualified for such predictions. EC is 3 because all of both physical evolution and biological evolution quantification apply.
31	YEC is 0 because there is no room for anything but inerrant biblical literalism. OEC is 1 and IDC is 2 because the disallowed spaces are not qualified for such observations and documentations. EC is 3 because all of both physical evolution and biological evolution quantification apply.
32	YEC is 0 because there is no room for anything but inerrant biblical literalism. OEC is 0 and IDC is 0 because predictions from completing observations in disallowed spaces are not considered. EC is 3 because all of both physical evolution and biological evolution quantification apply.

Perfect consistency with modern science would result in a total score of 96 (32 x 3 = 96). The four total scores for YEC, OEC, IDC, EC are 0, 22, 36, 92, respectively. This represents 0%, 25%, 42%, 96% of a total possible of score of 100%. These percentages are compared visually in Figure 24.

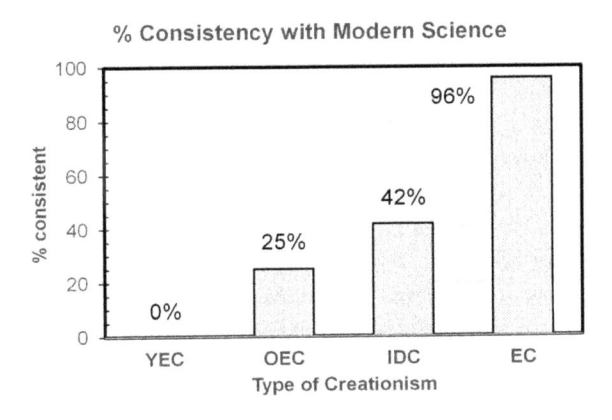

Figure 24. Consistency between modern science and each of the four positions.

DISCUSSION OF THE COMPARISONS

Figure 24 is a summary of Table 7, based on the justifications in Table 8. YEC is seen to be completely inconsistent with modern science for all of its thirty-two specific characteristics as summarized in Table 7. Therefore, YEC is not scientific with a score of 0%. According to these scores, acceptance of YEC requires a rejection of modern science. On the other hand, using this scoring method, EC is the most consistent with modern science with a score of 96%. OEC and IDC are each less consistent with modern science than EC (with scores of 25% and 42%) but more so than YEC. For those who will undoubtedly criticize our scoring and conclusions, we encourage them to construct their own specific comparison tables, score the four positions in some kind of side-by-side method, and publish the detailed results in the same spirit that scientific hypotheses are presented in the open marketplace of scientific discussion. Here we have summarized how our scoring was done in a spirit of unbiased fairness.

Accepted theories in modern science receive a 3 for all thirty-two rows for a total score of 96. See especially #17 for a "theory" or explanation that is so thoroughly tested and confirmed that it is held with great confidence. For example, the theory of evolution satisfies this definition. Except for #24, "Explanations can be falsified," and #25, "Falsified explanations are rejected in favor of ones that agree with all applicable observations," EC receives a 3 for all other rows. EC accepts the theory of evolution except for the assumption that God initiated evolution and guides its processes (genetic mutations that ultimately translate into the diversity of life). Also, from Table 6 in chapter 5 we see that EC accepts that Jesus Christ is the only path to salvation, which cannot be falsified since it is a spiritual belief. In addition, the EC explanation for God's role in evolution cannot be falsified, so that #24 must receive a score less than 3. All other rows are identical with science and receive a 3. Using this same reasoning, each of the other three positions (YEC, OEC, and IDC) receive a 0, 1, or 2.

This means that EC is consistent with modern science, except for this additional understanding about God's role and Jesus Christ as the only path to salvation. A strictly scientific understanding without a God hypothesis assumes that there are only natural materialistic causes for the initiation of life and that there is no divine guidance for evolutionary processes and that only random and unguided processes cause genetic mutations. A genetic mutation at the molecular level can be changed by material causes.[6] EC assumes that in addition to these material causes there is also a divine guidance that occurs in nature. This assumption falls outside the boundaries of methods associated with modern science. Acceptance of this EC position depends on a willingness to accept that there are other than material causes

---

6. A genetic mutation is a change in the DNA base pairs.

at work in nature, especially at the DNA level.[7] Therefore, it is not surprising that #24 and #25 are the only cells for a less than 3 score for EC.

Rows 1, 2, 3, 4, and 5 have to do with the overlap of certain groupings of phenomena with each other and their explanations. For example, the various phenomena concerning the movements of the sun, moon, stars, planets, comets, and apples falling from trees all overlap in the sense that their movements are all controlled by the force of gravity. The modern scientific view is that this is a natural or materialistic force versus a spiritual force. The YEC, OEC, and IDC positions all rely on a divine command theology for all creation type events, such as how the earth was created and sustained and species came into being. This puts all three positions in opposition with modern science, so that each gets a less than 3 for each of these five rows.

Rows 6, 7, and 8 relate to explanations of phenomena. Consider row 6, "Explanations must be supported by enough evidence," and IDC. This position accepts many of the explanations of modern science and the associated evidences. However, not all phenomena have explanations that are held by most of the experts. IDC proponents point to this occasional absence of universally accepted explanations as a weakness. Those explanations that do have enough evidence are apparently supported by IDC, so that for #8 and IDC we score that cell with a 2. YEC rejects all forms of evolution despite the existence of overwhelming empirical evidences in several scientific fields. OEC and IDC accept the modern science time line and physical evolution but not biological evolution.

The methods of good science (rows 21–32) were also scored in a similar manner. For example, in row 30, "Predictions from accepted explanations are compared to the data," IDC scores a 2 and EC scores a 3 because most (but not all for IDC) scientific explanations are accepted and compared to the data. As noted earlier in this chapter, the Inter-Academy Partnership (IAP) concluded that established scientific facts about evolution are rejected by the YEC supporters. For this reason, in row 30 and YEC is scored a 0.

The bar chart in Figure 24 visually shows the results of our scoring. EC has a percentage total score of 96%, indicating that EC is nearly completely consistent with modern science. Its only two scores less than 3 are due to the distinction between science and the belief in God's role in evolution and in Jesus Christ as the only path to salvation. Science deals only with material phenomena. A belief in some unspecified divine interaction with the material world would require belief in God, which is beyond the role of science. Therefore, these two scores cells should not be surprising.

---

7. McFaul and Brunsting, *God and Randomness*. The authors suggest that there is the possibility of additional dimensions, not accessible to us, but accessible to the divine.

From the bar chart (fig. 24) IDC has a 42% consistency with modern science. This results from IDC the underlying assumption that biological evolution must be rejected because it leads to atheism. Also, evolution cannot explain some of the irreducible complexities of nature. This view is not data-centric, subjected to peer review, and does not encourage comprehensive observations and testing.

YEC and OEC are quite similar with one basic difference: OEC accepts the age of the earth as 4.5 billion years old, using an interpretation of the Genesis creation account of a "day" as not being literally twenty-four hours or one thousand years. YEC interprets that account as a "day" equals one thousand years. Applying these two positions to the thirty-two specific characteristics of modern science, we get 0% and 25% to describe each of their consistencies with modern science.

CONCLUSIONS

Throughout this chapter, we have followed the National Academy of the Sciences' twenty characteristics (with twelve more added, for a total of thirty-two) to compare the four distinct positions that Christians hold on the relationship between Christianity and modern science. In tables 7 and 8, we assigned numerical scores that show how they differ from each other. The bar graph in Figure 24 summarizes the details that we have presented in these two tables. The most conservative of the four is Young Earth Creationism (YEC) that follows in full the doctrine of biblical inerrancy or literalism. The followers of YEC hold that God created the universe six thousand to ten thousand years ago along with all the species that inhabit the earth in six days through divine command.

For the other three positions of Old Earth Creationism (OEC), Intelligent Design Creationism (IDC), and Evolutionary Creationism (EC), the YEC view is too narrow and restrictive. Instead, they follow different approaches. Both OEC and IDC agree in part with modern science. They accept the big bang timeline and apply the theory of evolution to the development of the physical universe. However, they reject biological evolution and hold to the divine command doctrine of plant and animal creation. EC is the only one of the four position to accept the big bang timeline as well as both physical and biological evolution.

Each of these four positions emerged in response to the development of Christian fundamentalism that started in the early twentieth century. Over time, the proponents of these four positions developed different approaches to how Christianity should engage modern science. YEC opted for non-engagement, OEC and IDC for partial engagement, and EC for full engagement. Biblical literalism groups like YEC applied the doctrine of inerrancy to reject modern science. The OEC and IDC modifications led to partial acceptance, and the EC changes resulted in full acceptance.

Despite their disagreements over how to apply the doctrine of biblical inerrancy to science and evolution, there is one area where they are in complete agreement: The Bible is inerrant in its proclamation that the only pathway to eternal salvation goes through Jesus Christ. The Gospel of John 14:6 is often cited to support this belief. In this quotation, Jesus claims, "I am the way, and the truth, and the life. No one comes to the Father except through me." In his sermon as recorded in Act 4:11–12, Peter proclaims to the elders and scribes, "This Jesus is 'the stone that was rejected by you, the builders; it has become the cornerstone.' There is salvation in no one else, for there is no other name under heaven given among mortals by which we must be saved." Thus, on the subject of Christ as the only means of salvation in life after death, the followers of all four positions stand united.

As electronic communication and mass transportation systems drive the world in the direction of becoming a global village, the trend toward greater religious pluralism continues to grow. As a result, Christians find themselves increasingly in the presence of the adherents of other religions who do not share this belief. As we have seen thus far in our discussion of the relationship of Christianity and science, Christianity evolved from a position of complete rejection of the discoveries of modern science (YEC) to total acceptance (EC). Can we discern a changing pattern in how Christians relate to the members of other spiritual communities as the global village continues to grow? And what impact is the belief that the only pathway to salvation goes through Christ likely to have as Christians interact with greater frequency with the followers of other faiths? Starting with chapter 7, we begin addressing these questions.

7

Religious Pluralism

INTRODUCTION

The second part of this book centers on Christianity and religious pluralism and is an extension of the first part, in which we examined premodern science and modern science. After comparing the four Christianity and science positions (ch. 6) in relationship to NAS's modern science criteria, we concluded that Evolutionary Creationism satisfies these criteria more than the other three. Evolutionary Creationism is the only position to accept that the universe came into being with the big bang 13.8 billion years ago and expanded through a process of both physical and biological evolution that led to the development of self-conscious human life on earth.

While it might appear that there is no connection between the topics of Christianity and science and Christianity and religious pluralism, the opposite is true. They are closely connected. In our effort to overcome the time and space limitations of human contact and interaction, we have applied our modern scientific knowledge of digital electronics and jet engine technology to create a worldwide web of internet exchange and rapid air travel. The earth is becoming a growing global village. While our planet has not decreased in physical size as a result of these modern innovations, we are evolving into a different world compared to the communication and transportation patterns of premodern cultures that remained separated from each other by geographic barriers such as wide oceans and high mountains.

Because of modern scientific advancements, the world really has changed in many ways.

In the realm of religion, this has stimulated the increasing out-migration of individuals from one country to another. While our planet has always existed and still exists in a condition of religious pluralism, modern forms of communication and transportation have added a new layer of global interfaith interaction. No doubt, the older arrangement of pluralism and limited cross-cultural contact continues, but a newly evolving pattern of increased interaction has grown along with it, as many observers have noted.

For example, since 1991 the Pluralism Project at Harvard University, under Diana L. Eck's guidance, has tracked the growing trend toward religious pluralism in the United States.[1] When the 1965 Immigration Act eliminated international immigration quotas, it opened the door to greater religious diversity. Immigrants from formerly excluded regions of the world started migrating to the United States in unprecedented numbers. As a result, Muslims, Hindus, Buddhists, Sikhs, and others began arriving in the United States from all regions of the globe. This pattern of expanding religious pluralism was duplicated throughout Europe, Canada, and many other nations.

Other groups parallel the Harvard Pluralism Project and follow religious demographic shifts that occur throughout the entire planet. The International Religious Demography project that started in 2008 collects and analyzes statistical trends from various worldwide sources and publishes results in its World Religion database. The Center for the Study of Global Christianity possesses over one million documents related to Christian and religious demographic changes. The well-known Pew Research Center that was founded in 2004 publishes through the Pew Forum on Religion and Public Life updated information about changing worldwide religious demographics. In combination, these groups have described the growth of worldwide religious pluralism along with its potential impact on the nations and regions where it is occurring.[2]

As a consequence of evolving worldwide demographic trends, religious heterogeneity around the world is increasing. This is especially so for those societies with open borders that attract a diversity of individuals and groups from other countries and communities around the world. Even

---

1. Eck, *New Religious America.*

2. The International Religious Demographic Project is located at Boston University in Boston, Massachusetts; the Center for the Study of Global Christianity is found at Gordon-Conwell Theological Seminary, Hamilton, Massachusetts; the Pew Research Center is headquartered in Washington, DC. See Johnson and Grim, *World Religion Database*, and Johnson and Grim, *World's Religions in Figures.*

though some nations compared to others remain more closed to both in- and out-migration, the Internet allows everyone everywhere greater access to the larger world that exists outside of their own cultures. As a result, people do not have to leave their own geographical regions in order to learn about how others live in theirs. Through the worldwide web, everyone can witness the global evolutionary trend toward greater religious pluralism.

While we are increasingly aware of the degree of diversity that surrounds us, we should not lose sight of how recently this remarkable change came upon us. When we search for the back story that brought us to this point, we can identify Johannes Gutenberg's invention of the printing press in 1440 as a key moment when this significant shift began to take place not only in science as discussed in chapter 3, but also in the realm of religion. Before the printing press, general illiteracy was widespread among the population. After the printing press, literacy levels increased steadily in every European country and in other locations where it was introduced. As a result of this one invention, during the sixteenth and seventeenth centuries, church leaders from different religious communities during the Reformation began to translate the Bible into many diverse languages.

Once started, this led others to translate non-Christian scriptures into the English language in order to make the sacred writings of non-Christian faiths accessible for the first time. In 1785 Charles Wilkins produced the first English translation of the best-known Hindu sacred book, the Bhagavad Gita. The first English translation of the Buddhist Dhammapada appeared in 1840. In 1868, John Chalmers translated the first English version of the Taoist text, the Tao Te Ching.

Paralleling this initiative along with greater international mobility, leaders of the various world religions took a major evolutionary step forward in interfaith dialogue by meeting for the first time in human history. This ground-breaking event in religion akin to the Copernican revolution in science occurred in 1893 when they met together at the World's Fair: Columbian Exhibition in Chicago, Illinois, and created the Parliament of the World Religions. Building on the momentum, translations of other scriptures continued. In 1905, Abdul Hakeem Khan translated the Qur'an into the first English edition. By the end of the twentieth century and continuing into the twenty-first, virtually all of the scriptures of the world religions have been translated into multiple languages and available through electronic networks. The net result of this process is that we are now more aware than ever of the extraordinary diversity of the faith communities and religious ideas that populate our planet.

RELIGIOUS PLURALISM AS DEMOGRAPHIC

In Table 9 below, we compare the size of the world religions along with their percentage of membership in global population of about 7.5 billion. As the following statistical comparisons show, no single religion dominates the world with a membership level greater than 50%, although the vast majority of the world's population hold some form of religious affiliation. With 2.4 billion members (32%), Christianity is currently the world's largest religion. Islam ranks second with 1.8 billion followers (24%). Hinduism is third with 1.2 billion (16%). Number 4, Buddhism, claims 500 million members (7%). In combination, these four worldwide religions have 5.9 billion followers or 79% of the world's population. This means that nearly four out of five people in the world are either Christian, Muslim, Hindu, or Buddhist. In addition to the eleven religions that appear in Table 9, there are 1.2 billion people (16%) who are not affiliated with any religion.

Table 9. World Religions by Size and Percentage

Religion	Size	Percentage, %
Christianity	2.4 billion	32
Islam	1.8 billion	24
Hinduism	1.2 billion	16
Buddhism	500 million	7
Taoism/traditional Chinese	394 million	5.5
Sikhism	30 million	0.32
Judaism	14.5 million	0.2
Jainism	4.2 million	0.06
Shinto	4 million	0.06
Zoroastrianism	2.6 million	0.04

A world map of some of the major religions and where they are located predominantly is given in Figure 25. Generally, Christianity is found in most of North and South America, southern part of Africa, the northern part of Asia, and Australia. Islam exist as a major presence in much of the northern half of Africa and in much of eastern Africa, the Middle East, and Indonesia. Hinduism is the dominant religion of India. Buddhism and various local religions are located mostly in China and East and Southeast Asia. Judaism is the main religion of Israel.

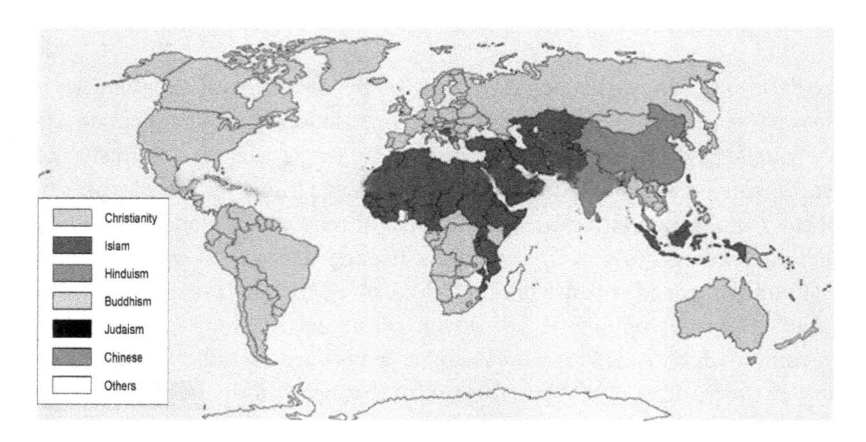

Figure 25. Worldwide distribution of some of the major religions.

As the twenty-first century goes forward, the religions of the world will continue to grow and disperse through the globe, but no single one of them will expand fast enough to exceed 50%. According to the religious research organization, the Pew Research Center, if current trends continue, by the year 2070 both Christians and Muslims will be equal in size at around three billion adherents each. After 2070, Islam will surpass Christianity in size because of higher fertility rates per family (3.1 for Islam and 2.7 for Christianity). Muslims are expected to be 34.9% of the world's population, and Christians 33.8%. Other religions will continue to grow as well, but none of them will be larger than either Islam or Christianity. Thus, when compared to the world population as a whole, for the foreseeable future all religions will continue to remain in the minority.[3] Religious pluralism is here to stay.

RELIGIOUS PLURALISM AS NORMATIVE

In addition to comparing the world religions according to their demographic size and percentage of the world population, there is a second way of defining the nature of religious pluralism. It is normative, which is to say that it involves identifying moral standards and behaviors that would reduce and possibly eliminate interfaith hatreds and hostilities. The second

3. Comparative statistical summaries vary according to how different research organizations analyze and track the demography of religious patterns and trends. For a comprehensive view of the demographic changes that are likely to occur among the world religions to the year 2050, including migration projections, see Pew Research Center, "Future of World Religions."

definition of pluralism builds on descriptive comparisons and is value centered. It presupposes the continuing existence of the diverse theological and philosophical perspectives and communities. The highest attainable normative aspiration and goal is peaceful coexistence. If only one religion existed in the world, there would be no need for a normative definition of religious pluralism.

However, as shown in Table 9 above, humanity speaks with many different spiritual voices and no doubt will continue to do so in the foreseeable future. The main challenge is for the followers of the world's diverse religions to evolve in the direction of living justly and peacefully together; and this remains the high point of interfaith relations when we define religious pluralism in normative terms. In chapters 10–12, we will examine how this might be achieved and the role that Christianity might play in it.

Thus, one of the main purposes of this second part of the book is to examine alternative pathways that the world's diverse spiritual communities can follow in order to evolve toward greater peaceful coexistence. As history has shown repeatedly, this is no easy task. The followers of diverse faiths have all too often clashed over disagreements that are connected directly to their religious beliefs. In extreme cases, many devout believers have chosen martyrdom rather than abandon their deeply held convictions in the face of life-threatening aggression and violence.

In order to move away from hatred and hostility, we will consider how religious pluralism can be viewed as a platform for increasing the evolutionary potential for interfaith harmony rather than decreasing it. We recognize that there is no guarantee that deepening our knowledge of religious pluralism will evolve automatically to greater inter-religious tolerance and cooperation. However, what is certain is that the lack of awareness will increase greater misunderstanding that in turn will strengthen the potential for more hatred and hostility.

FOUR COMMON CONCEPTS AND FIVE TYPES

In order to compare the beliefs that divide the religions of the world into separate faith communities, we start by reiterating the four common concepts that apply to all of them, which we described in chapter 1. They are materiality, spirituality, transcendence with or without intervention, and immanence. As we discussed above, materiality is the first layer that relates to the physical world that we experience through the five senses and that we understood through the methods of modern science (see ch. 6, Table 7). The second layer is spirituality that refers to a nonphysical level of reality that differs from materiality and that we cannot access through the five senses or the methods of modern science. The third concept is transcendence that signifies that this spiritual power, however defined, exists outside of materiality and is capable of intervening into the historical process. The fourth concept is immanence that means the spiritual power is embedded or hidden within materiality.

The next step involves showing how the four common concepts of materiality, spirituality, transcendence with or without intervention, and immanence apply to the wide variety of dissimilar spiritual communities that exist around the world. At first glance, this might appear to be an impossible task, but we will show that it is not. The reason for this is that despite their diversity the world religions can be grouped into five combinations or types: monotheism, pantheism, polytheism and henotheism, animism, and atheism and agnosticism.

Except for atheism and agnosticism, which either reject or remain uncertain about the existence of a spiritual dimension, the other four types can be classified as a form of theism. *Theos* is the Greek word for God. When *theos* is combined with *ism*, we get *theism*, or the study of God. All four forms of theism, that is, monotheism, pantheism, polytheism and henotheism, and animism are based on the belief that there exists some kind of second layer of spiritual reality (God, gods, or spirits) that transcends or is immanent within materiality. As stated earlier, atheism assumes that no second layer exists. Only the first layer of materiality does. Agnostics accept that the first layer is real but are uncertain about the second.

In the chapters that follow, we will show in detail how materiality, spirituality, transcendence, and immanence are applied differently to monotheism, pantheism, polytheism and henotheism, animism, and atheism and agnosticism.

UNIVERSAL RECIPROCITY NORM

Given the degree of diversity that separates the world religions from each other, do they share anything in common? The answer is yes, and it is called the *universal reciprocity norm*. Unlike the theological and philosophical differences that separate them from each other, it is through morality that they are bound together. Another way to say this is that they share common ethical core. It is ironic that one of the major pathways to cooperation and peaceful coexistence among the world religions goes through morality, whereas theological differences all too often divide them.

In order to avoid confusion, we begin our discussion of the universal reciprocity norm by defining terms that are used in many different ways— ethics, morality, and values. One definition refers to ethics as the theological or philosophical theory that justifies a particular value or rule of moral conduct. For example, the sixth commandment of the Ten Commandments that are listed in the book of Exodus (see 20:13) prohibits murder. This commandment is a specific moral principle that is based on the revelation God gave to Moses on Mount Sinai. God's revelation is the broad theological and ethical framework for justifying the specific moral prohibition against murder. A second way involves not making any distinction in these three terms but instead to use them interchangeably.

Our approach follows the first of these two alternatives. We believe that despite the diversity of the theological and philosophical views that separate spiritual communities from each other, the universal reciprocity norm cuts across all of them. It is humanity's most basic ethical norm that justifies specific moral principles or values that extend from it. The universal reciprocity norm also goes by the name of the golden rule. It is the shared ethical standard by which followers of all the world religions measure their specific codes of moral behavior.

As the authors have shown in a previous book,[4] the universal reciprocity norm appears in the South Asian religions of Hinduism, Buddhism, and Jainism. It is also included in the Middle Eastern faiths of Zoroastrianism and the Abrahamic threesome of Judaism, Christianity, and Islam. The same holds as well for the Central and East Asian traditions of Taoism, Confucianism, and Shinto. Both the Greek philosopher Plato and the Roman Stoic Seneca identified this norm as applying universally to all humanity, as have the leaders of indigenous spiritual groups in North America and Africa.

The norm, or golden rule, is easy to grasp. Do unto others as you would have them do unto you. Stated differently, if you want others to treat you with goodness, then treat them with goodness. If you do not want others to harm you, then do not harm them. In other words, do not do to others what you would not want them to do to you. Despite the doctrinal dissimilarities that divide the world religions from each other, the universal reciprocity norm or golden rule sits at the center of all of them.

In addition, this universal norm can be translated into specific virtues that all cultures respect and support, such as truthfulness, kindness, compassion, mercy, forgiveness, love and justice, caring for the needy, and so on.[5]

4 McFaul and Brunsting, *God Is Here to Stay*, 145–50.

5. For a detailed discussion of how multiple virtues overlap in the world religions, see McFaul, *Future of Peace and Justice*, ch. 8.

The quid pro quo formula for reciprocal behavior is straightforward: love for love, truth for truth, kindness for kindness, compassion for compassion, and so on. This means that the golden rule is both universal and reversible: what I would do to others, they would do to me. If I would harm them, they would harm me. If I want to be treated fairly, I should treat others fairly.[6]

Thus, the world religions are unified through the universal reciprocity norm; and because it permeates all of them, it can serve as a stimulus for increasing the potential for greater interfaith harmony. Can the diversity that exists between the world religions also serve this purpose? And if so, how? We will address these questions in chapters 10–12 after we have examined in chapters 8 and 9 how the different world religions combine materiality, spirituality, transcendence with or without intervention, and immanence into a diversity of patterns.

IN SEARCH OF THE ULTIMATE

In addition to the universal reciprocity norm that can be found in the world religions, is there any other dimension that overlaps them all? The answer is yes. It is the search for the Ultimate. While they differ in how they define and relate to the Ultimate, they are all united in the act of seeking it. How can this common quest serve as a foundation for evolving toward greater interfaith cooperation in spite of the theological and philosophical differences that arise as the world religions define it? We will answer this question in the final chapter.

CHRISTIANITY

One final topic remains. In the following chapters, we will give specific attention to how Christianity combines materiality and spirituality in relationship to both transcendence with or without intervention and immanence. We will also examine how Christian theology compares with the other world religions that integrate these four factors in different ways. In the process, we will take a close look at how Christianity can enhance the evolution of faith and foster greater interfaith harmony. At the same time, we recognize that the impact of Christianity does not go automatically in

6. The German philosopher Immanuel Kant (1724–1804 CE) called the universal reciprocity norm, or "golden rule," the universal moral law, which he defined as the "categorical imperative." For Kant, everyone has a duty to follow this moral law that he believed is valid intrinsically, good in itself, and should be followed in all situations. See Kant, *Metaphysics of Morals*.

one direction. All too often it has been a force for increasing hatred and hostility. Our goal is to identify how the Christian faith can be a force for evolving in the direction of greater good and not evil amid the religious pluralism that continues to expand throughout the emerging global village.

CONCLUSIONS

We began this chapter by describing the nature of religious pluralism in both demographic and normative terms. Our approach to dealing with the myriad of world faiths that populate the planet is to organize them into five types that include monotheism, pantheism, polytheism and henotheism, animism, and atheism and agnosticism. The dissimilarities that exist between these five types result from the many different ways in which they combine the four factors of materiality, spirituality, transcendence, and immanence.

Despite the diversity of theological or philosophical views that separate the world religions from each other, they share a common value called the universal reciprocity norm. It is also called the golden rule of doing good to others as you would want them to do good to you. Another version of the rule uses the negative form of not doing harm to others as you would not want them to do harm to you.

In addition, the followers of all the world religions are united in their quest to seek the Ultimate, even though they fall into different schools. In the process of discussing the similarities and differences that exist between the world's spiritual communities, we will examine how Christianity can stimulate the evolution of the global village toward greater interfaith harmony and away from increasing hatreds and hostilities.

We are now ready for chapter 8.

8

Monotheism

INTRODUCTION

In this chapter, we will concentrate on the five most important monotheistic religions that originated in the Middle East. They are Zoroastrianism; the three Abrahamic faiths of Judaism, Christianity, and Islam; and Baha'i. In the next chapter we will focus on pantheism along with animism, polytheism and henotheism, and atheism and agnosticism.

The following summary indicates the approximate time when each of these five religions emerged along with its main messenger and scripture.

Table 10. Summary of the Five Monotheistic Religions

Religion	Approximate Starting Date	Main Messenger (s)	Scripture
Zoroastrianism	1200 to 1500 BCE	Zoroaster	*Avesta*
Judaism	1300 to 2000 BCE	Moses	*Hebrew Bible*
Christianity	1 CE	Jesus	*Holy Bible*
Islam	622 CE	Muhammad	*Qur'an*
Baha'i	1844 CE	Bab and Baha'u'llah	*Book of Certitude*

As Table 10 makes clear, Zoroastrianism dates back to 1200 to 1500 BCE. Judaism is the first of the three Abrahamic faiths, which began between 1300 to 2000 BCE. Christianity is the second (1 CE), and Islam is third (622 CE). Baha'i, which is an offshoot of the Shi'ite branch of Islam, started in 1844 CE. Each religion has its most important messenger, and according to the above list they are Zoroaster, Moses, Jesus, Muhammad, and the Bab and Baha'u'llah. Each tradition has its own scripture: Avesta, Hebrew Bible (also called the Old Testament), Holy Bible, Qur'an, and the Book of Certitude.

COMMON ASSUMPTIONS

All five of these monotheistic religions share a common set of assumptions. They all accept that there exists a spiritual power called God who created the physical universe and the natural laws that sustain and direct its evolution. Although they presume the existence of a single Creator, they do not believe that this transcendent power is immanent within nature or the human body. God remains apart from it, as Figure 26 shows. The circle on the right is a visual image defined as God, which remains separate from the large circle on the left that is the human body.

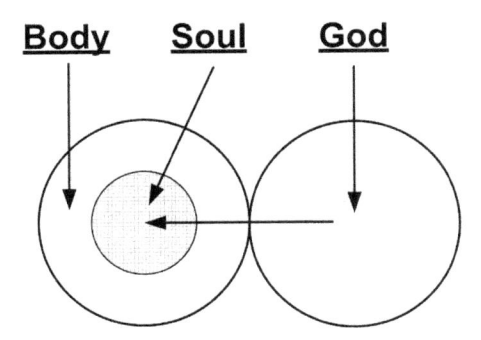

Figure 26. Diagram showing God's separateness from a human body.

As the large circle on the left shows, each human being has a separate soul (smaller circle) that God inserts into the body, although there is no agreement over whether this occurs at conception, at some point during pregnancy, or at birth. Plants, animals, material objects do not have souls. The vast majority of monotheists believe in life after death in the form of immortality of the soul or resurrection of the dead and that what happens to individuals after they die depends on the choices they made when they were alive. The followers of the monotheistic faiths believe in free will, which is a person's capacity to choose to be loyal or disloyal to the basic beliefs of their traditions. If they are faithful, their souls will be rewarded with a heavenly eternal life. If not, then their souls will go to hell. Each person gets only one life as a testing ground for what happens after death.[1]

For monotheism, the concept of revelation plays a central role. It is the means by which God bridges the gap between being the transcendent Other and finite humanity. God intervenes in history and chooses special mediators in specific historical settings and gives to them revelations or divine messages that indicate what God expects in the form of right beliefs and behavior. Eventually these revelations become written down and take the form of sacred scriptures that devout followers believe convey God's commandments and expectations, as well as the related rewards for obedience and punishments for disobedience.

The concept of sin holds a central place in all five of the monotheistic faiths. Sin means living by self-will rather than God's will; and avoiding sinful behavior is the key to living in right relationship with God. All of the monotheistic faiths hold that time is linear: At the time of creation, God set the universe in motion; and at some unknown point in the future, God will bring it to an end. In between the alpha and omega of God's time line, human life is a testing ground that determines each soul's eternal destiny in life after death.

In sum, the common concepts of (1) God as the transcendent Creator, (2) the existence of separate souls within each of us, (3) intervention through revelations given to chosen messengers and written down as sacred scripture, and (4) life after death in heaven or hell serve as a theological template, so to speak, for comparing the historic monotheistic faiths. In the remainder of this chapter, we will discuss all of them, starting with Zoroastrianism. We will examine their similarities and differences and identify connections that could help lay the foundation for greater peaceful coexistence among them.

ZOROASTRIANISM

As one of the world's oldest religions, dating back thousands of years, Zoroastrianism influenced the development of the other four monotheistic traditions. It was founded by Zoroaster who was also called Zarathustra in

---

1. In some of the religions that we will discuss in the next chapter, an individual gets more than one life time to attain the highest spiritual goal that is associated with that religion.

ancient Persia, now called Iran. For spiritual guidance, Zoroastrians turn to their sacred scripture the *Avesta* that contains Zoroaster's core revelations.

The Zoroastrian name for God is *Ahura Mazda*, the Creator, Sustainer, and Lord of the universe, which is a cosmic battleground for two great spirits. The first is the good spirit named *Spenta Mainyu*; the second is the evil spirt called *Angra Mainyu*. The struggle between these two opposing spirits pervades the entire universe. It is also that same micro battle that is being waged inside the conscience of every person. This struggle involves the challenges everyone faces when having to make either good or evil choices related to the countless number of moral dilemmas that all persons confront while they are alive. The three Abrahamic faiths of Judaism, Christianity, and Islam along with Baha'i integrated Zoroaster's image of the human struggle between good and evil into their theological worldviews.

This battle between good and evil will continue until God brings the universe to an end at some unknown point in the future, and goodness will triumph over evil. Then the Lord of the universe will judge all humans to determine their eternal fate. In the Zoroastrian worldview, each person's life is a testing ground, and what a person does during this time determines what will happen after death. God will judge everyone according to the relative balance of good or evil thoughts, words, and deeds as measured against the principles of truthfulness, charity, justice, and compassion. At its core, Zoroastrianism is a religion of ethical purity and action.

The Zoroastrian image of what happens in life after death includes two levels. The first occurs three days after the physical life of a person comes to an end, and the second on the final day of judgment. At the first level, individuals with a life pattern of predominantly good thoughts, words, and deeds will enter paradise. Where evil prevails over good, the soul is punished in hell. However, no soul remains in hell forever. Instead, the duration of punishment varies according to the severity of evil thoughts, words, and deeds. Some souls will remain in hell for longer periods of anguish and suffering than others, but eventually all souls will go to paradise because God's mercy is ultimately greater than God's justice.

JUDAISM AND FIVE COVENANTS

Judaism, which is the first of the three Abrahamic faiths, embodies the same monotheistic traits as Zoroastrianism. One of its unique features is its focus on the concept of covenant, which we define in order to avoid confusion. At the most general level, a covenant is an arrangement between two or more parties who agree to specific conditions that govern their relationship. This agreement involves acting in ways that are stipulated in the covenant. The words *covenant* and *contract* are often used interchangeably, but they are not identical. One of the main differences is that a contract is defined more narrowly as a legally binding agreement. A covenant might also incorporate the concept of legality, but it is not limited to elements that are enforceable by law. Covenant has a broader meaning that includes both legal and non-legal duties and obligations.

In the Abrahamic religions, the word covenant carries particular theological connotations that involve God's relationship to humanity as a whole as well as to Jews, Christians, and Muslims. It is a core concept that permeates the experiences of the ancient Israelites. Although disagreements exist over the number of covenants that appear in the Hebrew Scripture, for our purpose, we will identify five that are the most important.

The Old Testament begins with God's general covenant with Adam and Eve as recorded in Genesis 1:26–30 and 2:16–17. Described as the Adamic or Edenic covenant, the biblical account assures Adam and Eve that they can remain in the idyllic garden of Eden forever providing the do not eat of the fruit of the tree of the knowledge of good and evil. When they do, as symbolized by the apple, God marks them with the curse of sin and death and expels them from the garden. This begins the human sojourn through time along with the trials and tribulations of the generations that follow them.

The second covenant is called the Noahide covenant that is tied to the story of Noah and his family and to the great flood that destroys all the inhabitants of the earth because of human unbelief and wickedness (Gen 5:5–8). At the end of the flood, God creates a new covenant with Noah and his descendants. Through the sign of the rainbow, God promises Noah to never again destroy all the earth and its inhabitants through flood waters (Gen 9:13–17). For Jews, the Noahide covenant and its seven laws that apply to God's relationship to humanity as a whole. In addition, the Jewish Torah that consists of the first five books of the Hebrew scripture (Genesis, Exodus, Leviticus, Numbers, and Deuteronomy) identifies the 613 Mosaic laws that pertain specifically to the Jews.

After the Adamic and Noahide covenants, third on the list of the five covenants is God's covenant with Abraham, which is the most important unconditional covenant in the Hebrew Scripture. Starting in Gen 12:1–3, God blesses Abraham and promises to make his name great by turning his offspring into an exalted nation that will be a blessing to all the inhabitants of the earth. The Jewish belief that they have a special status as a chosen people whom God designated as an example of righteousness for all the world to follow is rooted in this covenant. For Jews, an essential aspect of the Abrahamic covenant is that God will never abandon them, because they are God's chosen people.

Number four is the Mosaic covenant that is perhaps the best known of all the Jewish covenants. According the Deut 30:1–10, it is a conditional covenant that involves both blessings and curses. The Mosaic covenant requirements are detailed in the Torah, as stated above, that contains the 613 laws that stipulate the beliefs and practices that the Jews were expected to follow. If they remained faithful to their covenant, God would bless them with a promised land of "milk and honey," which was the Canaanite territory that the Israelites conquered and inhabited.[2] If they violated their covenant, God would curse them and remove them from the land.

The books of Joshua through 2 Kings tell the story of the ancient Jewish rise to prominence and then fall into defeat and exile in Babylon. This story is written from the perspective of the Mosaic covenant for the express purpose of showing how the kings and community caused their own demise, because they violated the duties and obligations associated with their special Mosaic covenant with God. When they were obedient, God gave them a homeland; and when they were disobedient, God removed them from it. Despite the history of defeat and occupation by outsiders such as the Assyrians, Babylonians, Greeks, Seleucids, Romans, and persecution during nearly two thousand years in the Diaspora, the Jews never abandoned their belief that God chose them for a special purpose. When the United Nations created the modern state of Israel in 1948, many Jews saw this as God's abiding faithfulness. Once again, the chosen people were in possession of their historic homeland.

The fifth and last Old Testament covenant is tied to King David. The unconditional Davidic covenant is described in 2 Sam 7:8–16; 1 Chr 17:11–14; and 2 Chr 6:16, where God pledges that the kingdom of David will never end. For many modern Jews, the establishment of the modern state of Israel can be viewed through the lens of this fifth covenant. Thus, for the Jews, God intervened on five separate occasions from Adam through David to establish covenantal relationships with them and humanity as a whole.

In parallel with Zoroastrianism, Judaism includes all of the basic elements that comprise a theology of monotheism. There is one Creator God who intervened in history and selected messengers through whom God covenanted with the Jewish people and gave specific revelations that became an integral part of the Hebrew Scripture. There is one slight variation compared to the other forms of monotheism. Early Judaism did not include

2. Some writers consider God's promise to reward the ancient Israelites with their own homeland as a fifth covenant called the Palestinian covenant. We consider God's promise to give the ancient Jews a land of milk and honey part of the Mosaic covenant.

belief in the existence of separate souls. This appeared later after they encountered Zoroastrianism during the Babylonian exile. However, in lieu of
belief in separate souls and a spiritual afterlife, the ancient Jews held that
God would reward their obedience to the Mosaic covenant with a promised
homeland. For many current Jews, the creation of the modern state of Israel
in 1948 is a continuation of that promise.

CHRISTIANITY AND TWO COVENANTS

Unlike Zoroastrianism that stands outside of the Abrahamic traditions,
Christianity is the second religion in the Abrahamic sequence. Like its parent religion of Judaism, Christianity also is based on a covenantal theology.
Jesus's early followers did not view themselves as starting a new religion.
Instead, they believed that his life and ministry signified the culmination of
the Davidic covenant. They called him the Messiah or anointed one, which
meant he was heir to the kingdom of David that would never end. In the
process of converting the Jesus of history into the Christ of faith, the early
Christian movement called The Way reduced the five Jewish covenants to
only two—an old and a new covenant.

Evidence for this transition can be found in both the New Testament
and Old Testament that Christians held forecast the coming of Jesus as God's
Messiah. For New Testament writers, the book of Jeremiah (31:31–34) fore-
told the coming of a new covenant. "The days are surely coming, says the
Lord, when I will make a new covenant with the house of Israel and the
house of Judah. . . . I will put my law within them, and I will write it on their
hearts." Building on this Old Testament background, the New Testament
writers proclaimed through their four gospels and numerous letters that
God had created a new covenant in Christ.

The Apostle Paul, whose letters are the oldest writings in the New Testament and preceded the four Gospels, was the first to lay the foundation of
this new understanding. He writes in 2 Cor 3:6 that Christ has made his followers to be "competent to be ministers of a new covenant, not of letter but
of spirit." He reiterates this theme in Rom 7:6. "But now we are discharged
from the law, dead to that which held us captive, so that we are slaves not
under the old written code but in the new life of the Spirit." In 1 Cor 11:25,
he ties the new covenant to the words that Jesus spoke to his disciples at the
Last Supper. Jesus took a loaf of bread, gave thanks, and said: "'This is my
body that is for you. Do this in remembrance of me.' In the same way he
took the cup also . . . saying, 'This is the new covenant in my blood. Do this,
as often as you drink it, in remembrance of me.'"

When the New Testament Gospels were written starting about thirty years after Jesus's death, Mark was the first of the three Synoptic Gospels[3] to be written followed by Matthew and Luke. For all three of the Synoptic Gospel writers, the Davidic covenant is the most important of the Jewish covenants because Jesus's birth lineage descends through David's family tree in the tribe of Judah. In his gospel, Mark (14:22–24) reiterates Paul's earlier reference to the words Jesus uttered at his Last Supper. "Take, this is my body." Then he took the cup and said, "This is my blood of the new covenant, which is poured out for many." When the other two Synoptic Gospels of Luke and Matthew appeared about two decades after Mark, they included Mark's reference to Jesus's Last Supper along with the dialogue that took place between Jesus and the disciples. (See Matt 26: 26–28 and Luke 22:19–20.) In combination, Paul's letters and the three Synoptic Gospels share the same belief: In Christ, God created a new covenant.

Nowhere in the New Testament is the distinction between the old and new covenants drawn more sharply than in the letter to the Hebrews. In 8:6–7, the author writes: "But Jesus has now obtained a more excellent ministry, and to that degree he is the mediator of a better covenant, which has been enacted through better promises. For if that first covenant had been faultless, there would have been no need to look for a second one." In 9:15, the division between the two covenants is even stronger. Jesus "is the mediator of a new covenant, so that those who are called may receive the promised eternal inheritance, because a death has occurred that redeems them from the transgressions under the first covenant." Whereas the Old Jewish covenant of the law brought death, the new covenant in Christ brings eternal life.[4] The old covenant was based on works and the new covenant on grace.[5]

As in the case of Zoroastrianism and Judaism, Christianity follows the patterns of the monotheistic faiths. This includes belief in a single

3. The word *synoptic* means presenting or taking the same point of view. It is used especially with regard to the first three gospels in the New Testament.

4. Many conservative and literalist Christians divide the biblical time period and history of the church into several dispensations. While the number of dispensations varies from three to seven, they all culminate in the second coming of Christ and the creation of a New Jerusalem. The vast majority of Christians, however, interpret Jewish and Christian history using a multiple covenants method. For this reason, we will not discuss Christian Dispensationalism.

5. The Reformed church theological tradition includes a Trinity-based covenant of redemption, in which God and Christ agree to redeem an elect group of faithful followers. Because of Christ's obedience unto death, God establishes a new covenant in his name. The Holy Spirit does Christ's work through grace in the lives of the elect. For details, see Fesko, *Covenant of Redemption*. Because the covenant of redemption is limited mainly to the Reformed church, we are not including it in our discussion of multiple covenants.

transcendent Creator God and separate souls. For Jews and Christians, God intervenes in history through covenantal revelations that become written down as sacred scriptures—the Old and New Testaments. Christianity also incorporates ideas about the angelic forces of good and the demonic forces of evil and shares images of heaven and hell in life after death.

In addition to these similarities, Christianity goes beyond the mono-theistic framework. In all the other monotheistic faiths, God chooses mes-sengers or prophets whose main role is to deliver the revealed word of God. The messengers are secondary, and their revealed words are always primary. Not so for Jesus. Through the doctrine of the Trinity, Christianity holds that Jesus as the Christ is coequal with God and the Holy Spirit. None of the other monotheistic religions worships its prophetic messengers, but Chris-tianity worships Christ as the savior who died for the sins of the world. It is Christianity's core belief that any individual who accepts Christ as a personal savior will receive salvation in heaven after they die. This is the essence of the Christian concept of the new covenant, as expressed in John 3:16. "God so loved the world that he gave his only Son, so that everyone who believes in him may not perish but may have eternal life." As we have shown in chapter 5, Table 6, the followers of all four positions of Young Earth Creationism, Old Earth Creationism, Intelligent Design Creationism, and Evolutionary Creationism are unanimous in their belief that the only pathway to eternal salvation goes through Jesus Christ.

ISLAM AND ONE COVENANT

Following Judaism and Christianity, covenant theology is central to Islam's understanding of its relationship to God. At the same time, unlike the five covenants in Judaism or the two in Christianity, in Islam there is only one covenant that applies to God's relationship to the whole of humanity. Surah 7:172 makes this clear, "And [remember] when your Lord took from the Children of Adam—from their loins—their descendants, and made them speak out about themselves: 'Am I not your Lord?' They said: 'Yes! We tes-tify for sure.'" The view of the covenant that is conveyed in the Qur'an is that God is the transcendent Other who commands obedience. At the same time, God is the reciprocal Other who promises mutual obligations.

This means that while God expects submission to the revelations that were given to Muhammad and written down in the Qur'an, God also re-ciprocates with blessings. Surah 30:6 states, "Allah never departs from His Promise." To repentant sinners, God offers forgiveness. To faithful, just, and righteous persons, God promises rewards. Surahs 20.82 and 20:112 read as follows: "And verily, without doubt, I am [also] He Who forgives again and again, those who repent, believe, and do right." "And he who works acts of righteousness, and has faith, [he] will have no fear of harm nor of any stopping of what is due to him."

Thus, all three of the Abrahamic faiths are grounded in some form of covenantal theology through which God has intervened into history. Where they differ is in how many covenants God has made. For Judaism, there are five; for Christianity, two; and for Islam, only one. While Zoroastrianism is also a member of the family of monotheistic faiths, it is not grounded in covenantal theology. Nor does it trace its origin back to Abraham. At the same time, for the followers of Zoroastrianism, God—Ahura Mazda the Wise Lord of the universe—is the Creator whose righteous spirit permeates the good thoughts, words, and deeds of all devout disciples.

BAHA'I: DISPENSATIONS AND MANIFESTATIONS

Baha'i, which was founded by the Bab and Baha'u'llah in the nineteenth-century, grew out of the Shi'ite branch of Islam in Iran. It is the fifth major faith that falls within the broad monotheistic framework, but it does not follow a covenantal approach to theology. Instead, it uses a theology of progressive revelation that is based on the writing of Baha'i's two founders, especially those of Baha'u'llah.[6] We will describe what Baha'i means by the concept of progressive revelation and compare it with the theology of covenants. For the followers of Baha'i, progressive revelation refers to the belief that the will of God has become known "with greater clarity" over time as a result of intervening in history through a succession of revelations. God's revelations are not limited to Zoroastrianism and the Abrahamic faiths but includes Hinduism and Buddhism, as well.

The key to understanding the Baha'i view of progressive revelation lies in its interpretation of the phrase: with greater clarity. For Baha'i, new revelations supplement previous revelations. There is no perfect or final revelation, and later revelations do not replace earlier ones. This means that the Christian concept of the new covenant does not supplant the old covenant of Judaism. Nor do the revelations of the Qur'an supersede those of the Old and New Testaments. The words "with greater clarity" mean to "add to" the revelations that already exist.

For Baha'is, there is only one universal religion, and the historic religions of Zoroastrianism, Judaism, Christianity, Islam, Hinduism, and

6. For the Baha'i theology of progressive revelation, see Baha'u'llah, *Book of Certitude*.

Buddhism are separate expressions of it. All of them are of divine origin even though they emerged in different time periods that Baha'i calls dispensations. Each revelation is distinct and relevant to the specific spiritual needs of people who lived during the dispensation in which it arose. In each dispensation, God sent a main messenger called a manifestation who revealed God's will for that religious era. The major manifestations that God sent during the different dispensations are Zoroaster, Moses, Jesus, Muhammad, Krishna, and Buddha.

Also, each dispensation produced its own scripture or scriptures that contain the separate revelations that God gave to each manifestation or messenger during that time. All of the scriptures contain unique spiritual insights, and the combination of all of them is greater than any one of them. The belief that progressive revelations increase our knowledge of the will of God with greater clarity refers to the continuing accumulation of spiritual truths and not that later revelations improve upon or replace earlier ones. All revelations are unique and relevant to the dispensations in which they appeared. Together they bring greater clarity to the will of God.

The founders of Baha'i, the Bab and Baha'u'llah, are on the same footing as the prophets who preceded them: they are messengers of God. However, as a result of the Baha'i concept of time cycles, their role is different. Baha'i divides history into two time cycles. The first is called the Adamic Cycle and the second the Cycle of Fulfillment. The Adamic Cycle began with Adam and ended with Muhammad. It lasted about six thousand years. It was during this cycle that the different dispensations of Hinduism, Buddhism, Zoroastrianism, Judaism, Christianity, and Islam appeared along with their manifestations, specific revelations, and scriptures. The Adamic Cycle is over, and the Cycle of Fulfillment has begun.

The second cycle started with the revelations of the Bab and Baha'u'llah, whom Baha'is consider manifestations of the modern dispensation. In yet unknown future dispensations, God will send other messengers with additional revelations. The Cycle of Fulfillment will end when humanity has achieved its highest goal—the establishment of universal peace and justice on earth. Prior to this occurrence, Baha'i treats all the major scriptures as equal. There is no rank ordering of the sacred texts. Nor do later scriptures cancel out earlier ones. All are relevant to their times, and all teach us something about God's will with greater clarity.

The strength of Baha'i spirituality is that it takes religious pluralism seriously and preserves the best spiritual insights of the various religions without sliding into a hierarchical system that relegates some religions below others. Each revelation makes its own unique and progressive contribution. There is only one God, one humanity, and one universal religion of which the many separate faiths are distinct expressions of God's blueprint of progressive revelation that includes the two cycles.

Even though Baha'i and the Abrahamic religions use different words (covenants or dispensations) to identify distinct eras, along with Zoroastrianism they share the common belief that God intervenes into human history in different times and places. Where the Abrahamic traditions and Baha'i differ is in their views of the number of time periods (covenants or dispensations) through which God has acted as well as in the nature and meaning of that action.

For example, many modern-day Jews who are awaiting the coming of the Messiah tie this expectation to the fifth or Davidic covenant in the Old Testament. Christians proclaim that there are only two covenants, and the second or new covenant is final and supersedes the old covenant. For Islam there is only one covenant that God made with all of humanity, and Muhammad is God's final prophet. For Baha'i, Muhammad is the final prophet of only the Adamic Cycle that also includes Adam, Abraham, Moses, Jesus, Krishna, and Buddha. With the end of this cycle, God initiated a new Cycle of Fulfillment in which the Bab and Baha'u'llah are the modern-day manifestations or prophets. Thus, despite their language differences of progressive revelations or of one or more covenants, all four faiths of Baha'i, Judaism, Christianity, and Islam along with Zoroastrianism share a common theological framework that defines God as active in human history by intervening in different times and places.

In the next chapter, we will examine pantheism along with polytheism and henotheism, animism, and atheism and agnosticism. How do these religions differ from the monotheistic traditions we have described in this chapter? Can we apply a covenantal theology or a theology of progressive revelation to the non-monotheistic faiths as well as to the five we have described above? What role might Christianity play when the monotheistic and non-monotheistic faiths are combined? We turn to the next chapter to answer these questions.

9

Non-Monotheism

INTRODUCTION

In this chapter, we examine non-monotheistic religions starting with pantheism followed by polytheism and animism. We also discuss both atheism and agnosticism as alternatives to all forms of theism. Just as we used circle images in the prior chapter to visualize how the monotheistic Creator God is transcendent over the creation but not within it, we will repeat this process below starting with Hinduism. However, in order to avoid confusion, we begin with a broad definition of pantheism.

PANTHEISM

We define the word *pantheism* to mean that God permeates the universe and every animate being and inanimate object within it.[1] The following image helps us grasp how pantheists envision the relationship between spirituality and materiality. Imagine that someone is wearing an overcoat that extends from head to toe so completely that it is impossible to see who is underneath it. The person is covered completely and remains hidden within the overcoat. In comparing the overcoat image to pantheism, the concealed person is analogous to God (the spirituality presence) that is hidden (immanent)

---

1. *Collins English Dictionary*, s.v. "pantheism."

within the physical world (materiality). To complete the analogy, the same person who is concealed within the overcoat also made it and therefore stood outside of it (transcendent) during the process.

We will examine three of the most significant forms of pantheism: Hinduism, Taoism, and Sikhism. In combination, their core spiritual concepts differ from those of the five monotheistic faiths that we discussed in the last chapter. While both the monotheistic and non-monotheistic religions accept the existence of a single Ultimate, called God, they differ in their interpretation of the relationship of this Creator God to the creation.

Hinduism

Hinduism is the oldest and largest pantheistic religion in the world. It dates back more than five thousand years. It is the dominant religion of India and the world's third largest religion with nearly a billion-and-a-half members. It has no single founder. For centuries, a countless number of sages, both ancient and modern, contributed to the growth and evolution of this complex and multifaceted faith.

The Hindu name for God is Brahman, whose presence throughout the universe and human experience is interpreted in three distinct ways: monism, dualism, and polytheism. In this section, we will focus only on the monistic and dualistic forms. Later in this chapter, we will show how Hindus relate Brahman to their many polytheistic deities. As we examine this spiritual diversity, it is essential not lose sight of the core Hindu belief that there is only one Ultimate Reality: Brahman. The Hindu adage that there is only one Truth (Brahman) and many paths to it captures well this principle.

In Figure 27 we use circle images to visualize how Hindus combine the concepts of materiality, spirituality, transcendence, immanence to differentiate monism from dualism.

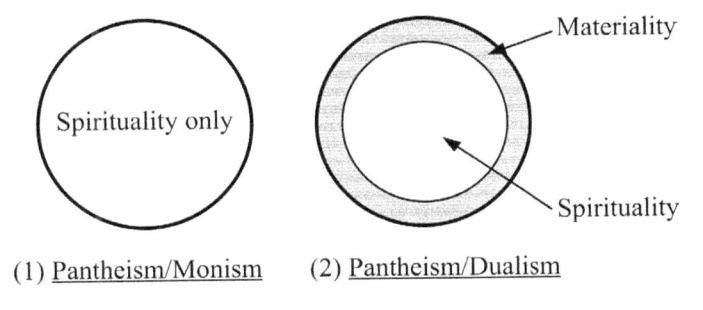

(1) <u>Pantheism/Monism</u>   (2) <u>Pantheism/Dualism</u>

Figure 27. Pantheism: Monism and Dualism.

The left-side circle refers to the monistic form of Hinduism in which only spirituality is perceived to be real. The right-side circle illustrates that in the dualistic form of pantheism both spirituality and materiality are real. How do we explain this difference given that both types of Hinduism accept that the transcendent God Brahman is immanent within the material universe? How can materiality be real in the right-side circle but not in the left-side circle?

Monists answer this question by starting with the assumption that only God is infinite and eternal. The physical world that we experience with our five senses is the body of God. Only God is permanent, and the natural environment is impermanent. Every animate and inanimate object in the material universe eventually ceases to be. As the left-side circle shows: only Brahman who is the spiritual Creator of the universe abides after every object in the material universe disappears.

To further illustrate their main point, monists turn to themselves and ask the following questions. Who is the real me? Is it who I am at this stage of my life? At my current age and physical appearance? Or is the real me who I was at an earlier age? Or is it who I will be later in my life if I live longer, even a minute from now? Is some part of my body real and other parts not real, such as my skin, bones, blood, cells, molecules, DNA, or other? In the midst of my every changing physical condition, can I identify any physical part of me that will not eventually perish? The answer is no. All materiality, including human biological existence, is finite and perishable. Only God is infinite and eternal.

What is clear in this discussion is that for monists the concept of reality is tied to permanency, whereas unreality is linked to impermanency. On this point, both monists and dualists agree. However, dualists define materiality differently than monists do. For monists, materiality is unreal, whereas for dualists materiality is penultimately real, which means next to real. Like monists, dualists hold that every aspect of the physical world is impermanent. Things change constantly except for God who is eternal and unchangeable.

However, dualists do not use this distinction to exclude the material universe from the realm of reality. Instead, they consider materiality to be a lesser or second level reality that stands below a higher spiritual reality in the same way that in the Army a general outranks a lieutenant general or in business an employer is superior to an employee. Both spirituality and materiality are real. The difference is that duelists view God or Brahman as Ultimate and the physical universe that is the body of Brahman as penultimate.

Hindus use the world Atman to describe the inner presence of the God Brahman who permeates every object, animal, or person in the universe.

Brahman is Atman, and Atman is Brahman. They are identical even though they have different names. Brahman is the macro manifestation of the divine spirit throughout the entire cosmos, and the Atman is the micro presence of Brahman throughout all the specific objects of the universe. Each Atman is of the same spiritual substance as every other Atman. It is this basic belief that qualifies Hinduism as a form of pantheism. The soul or Atman that permeates the cosmos in all its specific manifestations emanates from and is identical in its essence with Brahman.

The aim of the Hindu spiritual quest is to reach enlightenment, or *moksha*, and the pathway that leads to enlightenment is called reincarnation or transmigration. Hindus assume that through the process of reincarnation the hidden Atman is reborn into a sequence of different life forms before the moment of enlightenment occurs. Once this happens, upon the death of the person who has experienced *moksha*, the invisible Atman's transmigration journey stops. Hindus envision that the return of the Atman to Brahman is like a drop of water that blends back into the ocean.

Eventually, every Atman returns to Brahman; and when this occurs, the cosmic cycle that Brahman began unknown eons ago comes to an end. Then, through a process called eternal recurrence, Brahman starts another cosmic cycle. This new cycle is merely one among many in an unending number of cycles that the eternal Brahman creates repeatedly; and because Brahman is spiritually infinite, there is no end to the number of finite material worlds that Brahman can create.[2]

The pathway to enlightenment consists of right beliefs and behaviors. Right belief begins with the simple declaration "I am That" or "Thou art That" (*tat twam asi*, a Sanskrit language saying). Anyone who speaks this phrase recognizes that Ultimate Reality is not materiality but rather spirituality and that the transcendent God Brahman, who is also immanent throughout the entire universe, is the same as the spiritual Atman that is veiled within the self. Any person who believes that materiality is ultimate reality lives under a condition called *maya* or an illusion caused by confusion over the nature of Ultimate Reality. This can be corrected only when the individual accepts that infinite spirituality and not finite materiality is the true ultimate that permeates the cosmos and that lies hidden at the core of each person's being.

The first step of right belief is followed by the second level of right behavior called *dharma* or duty; and a person's performance of duty results in *karma*, which Hindus define as the law of moral conservation. Good

2. Two of the best-known primary scriptures in which the Hindu pantheistic understanding of God is expressed are the Bhagavad Gita and the Upanishads.

dharma leads to good karma and a higher reincarnation, whereas bad karma results in a lower reincarnation and slows down progress toward enlightenment. Thus, it is the combination of both right beliefs and behavior that guarantees progress toward the ultimate spiritual goal of releasing the Atman from rebirth.

In order to achieve this goal, Hinduism provides believers with multiple pathways. The saying "There is only one Truth (Brahman) and many paths to it" indicates that there are mainly four different types of dharmas or duties that a devout follower can pursue to reach enlightenment. Hindus describe the process of committing to one or more of the dutiful pathways as yoking oneself to dharma, which is synonymous with the word *yoga*. Another way to say this is that the four dharmas are the four forms of *yoga* that lead to the spiritual summit of *moksha*.

The four *yoga* pathways coincide with four different human tendencies or capacities: work, intelligence, emotion, and dedication. The first is called *karma yoga* that means the way of works. It refers specifically to following the duties of the caste into which a devout believer is born along with corresponding stages of the life cycle. The second is *jnana yoga* that focuses on developing the mind or intellect through the study of Hindu scriptures and sacred texts. Third is *raja yoga* based on meditation. The fourth is *bhakti yoga* that centers on devotion related to one or more of Hinduism's many polytheistic idols that we will discuss later in this chapter.

Whatever the combination of *yoga* options that Hindus choose to transmigrate through successive life forms on the journey toward enlightenment, believers assume that the proper performance of duty will lead eventually to ending rebirth. Thus, whether Hindus define their pantheism in terms of monism or dualism, the combination of ideas that include reincarnation, karma, dharma, and enlightenment give devoted followers a spiritual foundation of personal meaning and social stability. The other two major religions that share many of Hinduism's pantheistic concepts are Taoism and Sikhism.

Taoism

Taoism (also written as Daoism because the *t* is pronounced like a *d*) is one of China's two historically indigenous philosophies that emerged in the sixth century BCE.[3] It can be traced back to the warring states period that existed in ancient China from the eighth through the sixth centuries BCE. It was an

---

3. Confucianism is ancient China's other major religion that emerged around 600 BCE. We will not discuss it here, because its focus is mainly ethical and not spiritual.

especially violent time when fiefdom conflicts seemed endless and waves of aspiring warlords attacked each other in merry-go-round power struggles.

In their search to find ways to end the carnage they witnessed all around them, Chinese philosophers such as Lao Tzu, the founder of Taoism, turned to their ancient traditions for guidance. They started with the perception that there is an underlying order to the cosmos as a whole, an ultimate reality called ch'i (pronounced: chee) that encompasses the entire universe including the mental states of individuals. The ch'i is also called the Tao that can be defined as the breath of life or fundamental source that gives rise to all animate and inanimate objects, including humans.

The Tao is comprised of the two complementary opposites called the yin and yang as depicted in Figure 28.

Figure 28. Yin and Yang: Asian Religious Symbols.

The yin, or female forces, are dark, passive, soft, cold, wet, negative, and contracting. The yang or male forces are the opposite. They are bright, active, hard, warm, dry, positive, and expansive. The interplay between these two forces determines the extent to which nature, society, and the individual remain in a condition of harmony and calm or deterioration.

The yin and yang forces show up in life in countless combination: right-left, up-down, in-out, sound-silence, sweet-sour, birth-death, man-woman, and so on. This means that there will be harmony in the universe, society, and the individual, as long and the yin and yang remain balanced. The more that the dynamic between them shifts away from complementarity and toward conflict, the greater is the disharmony that appears within the self, society, and physical world. The goal, therefore, is to strive for maximum yin-yang stability and constancy.

Like Hinduism, Taoism starts with the assumption that an eternal spiritual force exists at the center of materiality. This force is called the Tao, and it consists of both the yin and the yang complementary opposites. However, from the Taoist's perspective, language cannot grasp the essence of the Tao, because it is ineffable. The following quotations from the best-known

Taoist scripture, the *Tao te Ching*, allegedly written by the founder of Taoism, Lao Tzu (or Laozi—Old Master), make this clear. "The Tao that can be told is not the eternal Tao." "If you try to hold it you will lose it." "The Tao is forever undefined." "The Tao is hidden and without name."[4]

At the same time, the indescribable Tao is the spiritual source of the material universe. It is the great mother of ten thousand things, which means that it is the origin of every entity. It has no beginning or ending. It is the infinite. "Tao is the source of the ten thousand things." "The beginning of the universe is the mother of all things." "The valley spirit never dies; it is the woman, primal mother." "The ten thousand things rise and fall without ceasing." "The Tao alone nourishes and brings everything to fulfillment." "The Tao lasts forever."[5] It gives birth to everything; and in death or disappearance, everything returns back into it. "Being at one with the Tao is eternal. And though the body dies, the Tao will never pass away."[6]

Although the spiritual source of all material existence cannot be grasped through language and is indefinable, nonetheless, Taoism uses many paradoxical images to describe it: hidden but present, full but empty, eternal and temporal, bottomless, origin of all, continuous, lives forever, invisible, inaudible, the subtle, infinite, boundless, without shape or form, vague, elusive, does nothing but produces everything, weakness that leads to strength, softness that gives rise to hardness.[7]

Living in harmony with the Tao leads to peace in self and society. Right beliefs should be followed by right behavior. This takes the form of non-action called *wu wei* that is based on imitating nature. It is not through goal-setting activities that we humans find peace and tranquility. It is the opposite of letting nature take its course. For Taoists, the image of flowing water captures the essence of behavior that takes the form of non-action or actionless action. When water flows down a hill, it does so without effort. Eventually it wears down everything in its path. Even the most rough-edged rocks yield to the steady flow of water that transforms them into smooth stones.

Like water, we too should go with the flow of life, which is what Taoists mean by non-action. Once again, we find ourselves encountering paradoxes: move but do not speak, live but do not possess, act but do not presume to know the outcome, live in the now, accomplish but take no credit, do not seek to change the world through forced activism but by non-action, keep

4. Feng and Jane English, *Tao te Ching*. The valley spirit and primal mother images refer to the indescribable spiritual center of the cosmos. These sayings are found in sections 1, 6, 29, 32.

5. *Tao te Ching*, sections 2, 8, 25, 34, 40, 41, 52, 62.

6. *Tao te Ching*, section 16.

7. *Tao te Ching*, sections 4, 10, 21, 22, 35.

your mind deep, live in goodness, honesty, and kindness, do what is right, do not contend, do not harm self or others, excel without the thought of mastery, be an adult by returning to infancy, do not desire greatness and you will become great, serve without serving, act without acting, taste without tasting, repay hatred with love, take small actions and they will lead to big outcomes, do not accumulate things but live simply, and when you know that enough is enough you will always have enough.[8] Thus, following the pathway of *wu wei* leads to harmony with the Tao, which in turn results in inner contentment and by extension outward to social peace.

Sikhism

Sikhism is the youngest of the three pantheistic religions. It began in South Asia, now India, during the sixteenth century CE when the two dominant religions were Hinduism and Islam. The origin of Sikhism can be traced to the storied account of how its founder guru Nanak (1469–1538 CE) combined many of the elements of both of these religions into a new pantheistic spirituality.

At age thirty, Nanak was swimming with a group of friends and suddenly disappeared. Although his friends searched frantically to find him, it was to no avail. They assumed he had drowned. To their surprise, he reappeared three days later; and when he saw them, he uttered a new saying: There is no Hindu, and there is no Muslim. This experience marked the beginning of Nanak's syncretistic vision that it was possible to combine elements from both pantheism and monotheism, despite their theological differences. As a result, he joined the ancient South Asian tradition of becoming a wandering guru who gathered around him a growing group of converts from both Hinduism and Islam.

When he died, Muslims and Hindus debated how to handle him in death because their traditions differed: be burned on the pyre or buried in the ground. He was laid out, flowers were placed on both sides—one for Muslims, one for Hindus. In the morning, whichever flowers still bloomed would determine how to deal with his deceased body. When his followers removed the sheet that covered him, he was gone, which symbolized that a new faith had been born.

The core of Nanak's new spiritual message centered on singing the ninety-nine names of God found in Islam's sacred scripture, the Qur'an; these names also appear in the Sikh scripture called the Adi Granth. He accepted the Muslim view that polytheistic idols do not exist and that there

8. *Tao te Ching*, sections 8, 10, 13, 19, 22, 28, 46, 51, 76.

is only one God, whom he called *Wehaguru*. At the same time, he embraced the Hindu form of dualistic pantheism that God is hidden within the material universe, which he assumed was a wall of falsehood also called the veil of *maya*. Each person has a dual nature that is both material and spiritual. Both are real. As in the case of Hinduism, a devout follower's goal is to penetrate through God's material body in order to merge with the spiritual core that is concealed within it.

In addition, Nanak accepted the Hindu belief in reincarnation but rejected the caste system. He assumed that when a believer's inner spirit merged with God, future transmigrations to new bodily forms would cease. The soul would blend back into God. Nanak also held to the Hindu doctrine of karma. Bad karma leads to continual rebirth, and good karma leads to progress toward spiritual fulfillment and eventual return to God.

Accumulating bad karma occurs when an individual seeks only to satisfy the ego through self-gratification, especially in the form of greed and attachment to worldly desires or success. Idolatry is identified as the worship of wealth. Accumulating good karma results from experiencing the presence of God by intentionally seeking spiritual goals. This includes reading the Sikh scripture, Adi Granth, and following its spiritual and moral teachings. The weekly community gatherings in a religious center called a *gurdwara* (the Sikh name for temple) consist of chanting the ninety-nine names of God and feeling God's inner presence. Accruing good karma also results from living with kindness and compassion toward others, being charitable by sharing goods with the poor, and participating in the weekly public meal offering called a *langar* following worship.

Thus, when we compare the three pantheistic religions of Hinduism, Taoism, and Sikhism, it is only the monistic form of Hinduism that does not accept the reality of materiality, whereas dualistic Hinduism, Taoism, and Sikhism do. Along with the monotheistic views discussed in the previous chapter, the pantheistic religions accept the existence of the soul although they interpret it differently as well as what happens to it after death. Monotheists hold that God determines the soul's eternal destiny after only one life, whereas Hinduism and Sikhism include the belief that the soul is reincarnated into new life forms according to a persons' good or bad karma until the individual becomes enlightened after which the cycle of rebirth ends. Nor do the monotheistic concepts of specific covenants or historical time cycles apply to the pantheistic traditions. Despite some areas of overlap, when we shift from monotheism to pantheism, we move into spiritual communities with vastly different worldviews. Can the same can be said for polytheism and henotheism?

POLYTHEISM AND HENOTHEISM

We define polytheism to mean belief in many gods and henotheism as a form of polytheism that places one of the many deities in a position of superiority over the others. Figure 29 shows this relationship. The large outer egg-shaped circle represents the material world of animate and inanimate objects. The small inner circles symbolize the spiritual presence of the polytheistic gods who inhabit the world. The largest circle stands for the major deity, and the arrows depict this deity's dominance over the lesser ones.

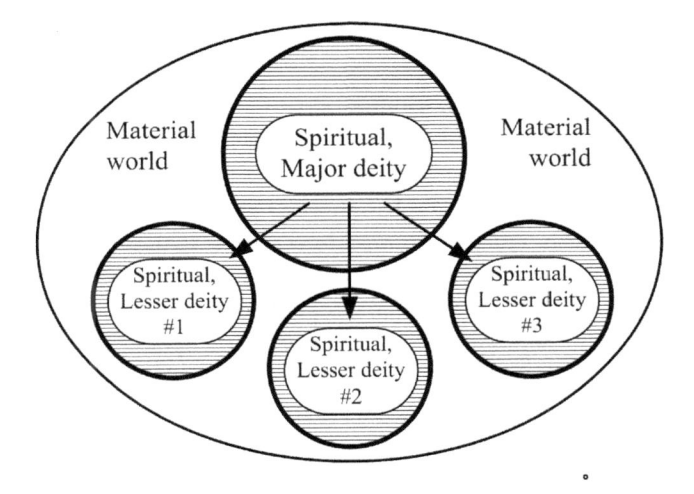

Figure 29. Polytheism and Henotheism.

The ancient Greek pantheon, in which the mighty Zeus dominated the other gods, is one of the best-known examples of henotheism. When the Romans conquered the Hellenistic world and beyond, they appropriated the full range of Greek deities and renamed them. The powerful Zeus became Jupiter, the god of the heavens and the sky, who stood above the rest of them as their leader. The henotheistic pattern of ranking multiple deities according to their divine status and function also appears frequently among the indigenous religions that inhabit the Americas, Africa, Asia, and Australia.

The largest polytheistic religion in the world is Hinduism. Given that we identified Hinduism as a form of pantheism earlier in this chapter, it might sound like a contradiction. However, it is not for one primary reason. After thousands of years of ongoing refinement and revisions of countless sages and gurus, Hinduism combines numerous spiritual traditions. Through the course of Hinduism's evolution as the world's third largest religion, one of the main challenges that its sages confronted was to reconcile the belief in a single God with the worship of multiple gods. In simple terms, it involved reconciling the one God Brahman with the many deities that Hindus worship.

It is through the concept of avatar that Hindus bridge the gap between the One and the many. An avatar is an earthly manifestation of Brahman. All of the various avatars or gods are emanations of the one God, including those who have a descending list of both male and female avatars of their own. Unlike the ancient Greeks and Romans who believed their deities were real, Hindu avatars have no autonomous existence. No matter where they rank on the avatar hierarchies or descend from higher gods, they all point beyond themselves to Brahman.

It is through the devotional form of Hinduism called *bhakti yoga* that Hindus worship one or more of the many avatars. The most popular among the males are Shiva (creator/destroyer), Vishnu (preserver) and his avatars Krishna and Rama, and the elephant headed boy Ganesh. The popular female deities include Kali and Durga, along with countless others. Numerous public festival and rituals are associated with the worship of Hinduism's large number of deities. When compared to other traditions, Hindus hold more indoor and outdoor religious festivals and ongoing worship observances than the followers of any other religion.

The following story captures how Hindus thinks about the relationship between Brahman and the vast number of avatars that emanate from him. At age twenty, a young man set out on a lifelong journey to determine the exact number of Hindu gods. As he traveled from village to village, he wrote all their names in his notebooks. When he reached the age of eighty, he had compiled scores of notebooks filled with deity names. Then, he counted them all, drew a line underneath the last name, and wrote down the final number: 1. He concluded that only Brahman is God, and all the rest are pathways to Brahman.

To sum, Hinduism is a pantheistic religion that consists of three types: monism, dualism and polytheism. The Ultimate is Brahman/Atman. The goal is Enlightenment (returning the Atman to Brahman). Reincarnation is the means to Enlightenment. Doing good dharma leads to good karma, which in turn accelerates progress toward Enlightenment that ends the reincarnation cycle. Amid the many spiritual pathways that devout Hindus may follow, in the final analysis they are united in the belief that there is but One Truth and many paths to it.

ANIMISM

Next, we focus on animism as another expression of the non-monotheistic religions. Animism shares much in common with polytheism, but it differs in one important regard. Polytheists use numerous statues and icon images to portray their many gods as external realities that have an identifiable presence in the world. Normally, animists do not follow this procedure, but instead hold that the animate and inanimate objects of the material world contain spirits that cannot be seen with the naked eye, as Figure 30 shows. The outer egg-like circle symbolizes the material world. The two middle-size, inner ovals represent the animate and inanimate objects that exist in the world. The two smallest ovals signify the presence of separate spiritual forces within the objects.

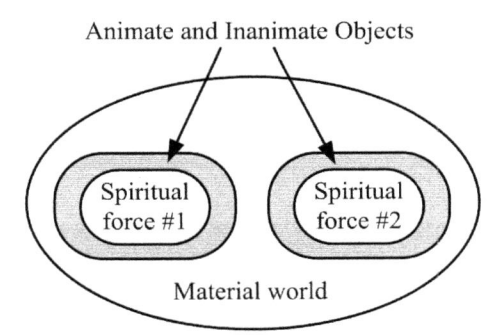

Figure 30. Animism.

When animists convert the hidden spirits into visual forms through carvings or other artistic renderings, they combine animism and polytheism or henotheism as in the case of the African masks or the totem images that North American indigenous tribes create.

Earlier in this chapter, we discussed the pantheistic belief that a single spiritual Ultimate Reality is immanent within the material universe. Animism differs from pantheism by assuming that material objects have separate spirits hidden within them rather than just one. While there exist numerous examples of animistic religions, one of the best-known, Shinto, can be found among the major spiritual traditions of Asia, particularly in Japan where it originated.

Like the other ancient religions of Central and East Asia, the worldview of Shinto (also called *shen tao*) is based on the yin/yang distinction that we discussed earlier under Taoism. In the case of Shinto, the Tao as the source of life includes and applies to the concept of the *kami*. The word kami refers to an inner spiritual presence in any person, place, or thing that inspires awe, respect, reverence. This involves both the both animate and inanimate objects in the material world, national heroes and leaders, and sacred spaces and places, even though there is no way to account for the total number of kami spirits that exist throughout the world.

Shinto traces its beginnings back to the sixth century BCE. The Shinto sacred scriptures the *Kojiki* and *Nihongi* written in the eighth century CE describe how two great kami spirits brought the world into being and how it developed. The first is the male kami, called Izanagi, and the second is the female, named Izanami. When Izanagi dipped his spear into the ocean (as visualized in Figure 31), the drips of water that fell from the spear tip created the Japanese Islands.

Figure 31. Creation of the Japanese Islands.

Then, out of the left eye socket of the male kami, Izanagi, came the Sun Goddess Amaturasu, the principle Japanese deity who protects the Japanese people. The large sun that sits at the center of the Japanese flag symbolizes the protective presence of the Sun Goddess Amaturasu. See Figure 32.

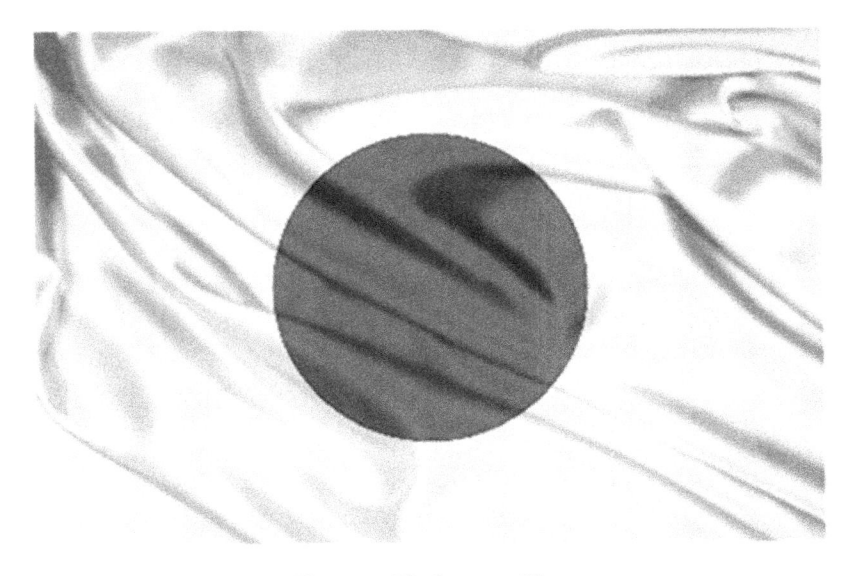

Figure 32. The Japanese Flag.

Amaturasu is the ancestor of Jimmu, the founder of the imperial dynasty in 660 BCE. Amaturasu also chose the Japanese people to be superior to all others and that one day they would rule the world. The Shinto kami creation story also claims that the Japanese islands sit at the center of creation. Between 1868 and 1945, these beliefs crystallized into the ideology of State Shinto that motivated Japan's military expansion throughout East Asia and the Pacific Islands before and during World War II.

After the War ended in 1945, militant State Shinto became transformed into today's peace-oriented, non-militant Shrine Shinto where followers continue to venerate the two great kami spirits Izanagi and Izanami that created the Japanese Islands and people. They also revere the countless other lesser kami spirits that dwell within nature, society, and individuals. Instead of pursuing the goal of political dominance and military conquest, Shrine Shinto exists to promote inner purity through right actions and relationships with the kami.

Spiritual pollution for self and society emerges when a person shows disrespect for the kami; and this occurs through attitudes and actions that involve greed, egotism, disregard for the nation, family, and other social groups, and by viewing nature as spiritless. The result is disharmony and social conflict. Restoration involves performing Shinto purification rites and rituals and by returning to a right relationship with the kami. Upon giving proper deference to the manifold spirits that inhabit and animate sacred spaces, places, nature, and society, Shinto's devoted followers look forward to the future when the original peaceful balance that was shattered through human selfishness will once again be re-established.

ATHEISM AND AGNOSTICISM

Along with the various forms of theism (monotheism, pantheism, polytheism and henotheism, and animism) that we have discussed in this and the previous chapter, we are also including a section on atheism and agnosticism. We start by noting that there exist two distinct types of atheism, and it is crucial that we keep them separate. The first is Western non-spiritual atheism, and the second is Eastern spiritual atheism. While this might sound like a semantic difference without a real distinction, this is not the case. Non-spiritual atheism relates to Western culture, whereas spiritual atheism applies to Eastern culture.

For Western atheists, only materiality exists, as shown in Figure 33 where the grey appears inside the circle. Spiritual forces that transcend nature or are immanent within it have no reality.

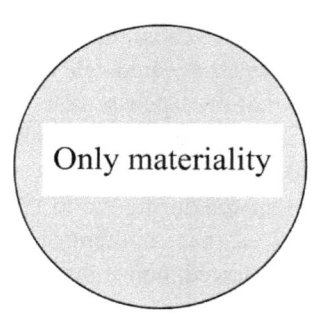

Figure 33. Western Atheism.

Some Western atheists are called hard atheists because not only do they reject belief in the existence of any kind of separate spiritual power, they are certain that their belief is true. Soft atheists also accept that spiritual powers do not exist but leave open the possibility that they might. Agnostics are like soft atheists. They are uncertain one way or the other about the reality of any spiritual presence that is greater than nature; and like atheists, they vary from hard to soft beliefs according to the degree of their certainty or uncertainty.

Western forms of atheism are typically directed against the monotheistic belief that there exists a supernatural, spiritual being that created the cosmos, directed its development, and performs miracles that violate the laws of nature. Passages such as Moses hearing a voice from a burning bush (Judaism), Jesus changing water into wine (Christianity), and Muhammad listening to the voice of the angel Gabriel to recite revelations that appear in the Qur'an (Islam) are such examples. These monotheistic stories are designed to show God's power over the material world and intervention into human history; and they demonstrate that God can intervene whenever, wherever, and however God chooses.

This is the view of a monotheistic God that Western atheists reject; it is consistent with their belief that there does not exist any kind of spiritual reality (God or gods) that transcends nature or is immanent within it. As shown in Figure 33, nature is the ultimate reality beyond which no other higher or greater realities exist. Building on this assumption, it follows that people delude themselves when they believe that some form of spirituality exists beyond materiality. Simply stated, the belief in God or gods is a false belief.

Furthermore, because Western atheists assume that materiality is ultimate reality, then any form of spiritual expression that emanates from belief in God or gods can be explained through some form of non-religious reductionism. For example, Sigmund Freud maintained that religious beliefs consist of psychological projections based on an infantile, neurotic need for a loving parent.[9] Karl Marx held that religion, the opiate of the masses, could be reduced to pecuniary factors that enable the dominant economic class in society to dupe, control, and exploit the subordinate class.[10] Emile Durkheim reduced religious beliefs to the collective conscience that contributes to social cohesion.[11] In all three of these atheist writers, nothing spiritual is real.

This is not the case for Eastern forms of atheism, for which attaining spiritual goals is of ultimate importance. Two of the foremost expressions of spiritual atheism are Buddhism and Jainism that emerged in South Asia, now India, during the seventh century BCE. Unlike Western atheism that rejects spiritual aims as delusional, the followers of Buddhism and Jainism hold that achieving the spiritual goal of nirvana is the main purpose of human life.

Buddhism

The Buddhist quest for nirvana is best seen against the backdrop of the Buddha's personal experiences. His real name was Siddhartha Gautama.

9. Freud, *Future of an Illusion* and *Moses and Monotheism*.

10. Marx, *Marx on Religion*.

11. Durkheim, *Elementary Forms of the Religious Life*.

His father was the nobleman who ruled part of northern India and kept Siddhartha behind the high walls of his palatial estate. The father's objective was to keep his son from seeing the surrounding poverty conditions and the destitute people whom he would govern. As he matured, Siddhartha grew impatient with these restrictions. In his early twenties, he ventured beyond the walls of his confinement to observe for the first-time what life was like on the outside.

During the brief outing away from his father's watchful eye, he witnessed four sights that changed his life forever and that served as the inspiration for his future teachings. He saw a diseased person, an old person, a dead person, and a group of wandering world-renouncing Hindu monks who were seeking Enlightenment. Upon observing the conditions of disease, old age, and death, Siddhartha concluded that the basic reality of life is suffering that results by being born with a biological body that deteriorates over time. Furthermore, he reasoned that reincarnation only perpetuates this condition from one form of life to another.

The fourth sight inspired him to seek his own Enlightenment in order to end as quickly as possible the endless cycle of suffering. Consequently, he abandoned his protected life of luxury and joined the wandering monks in order to master the spiritual practices that would lead him to moksha. After several years of seeking his Enlightenment, during which time he almost starved to death, he concluded that he was no closer to his goal than when he started.

Despite the feeling of discouragement, Siddhartha did not retreat back to his earlier life of privileged seclusion. Instead, he sat down under a fig tree, later called the Bodhi or Enlightenment tree, and vowed to meditate until he either died or achieved his goal of Enlightenment. It was during this time of intense meditation that he finally reached the end he was seeking: the experience of nirvana and the emotional certainty that he would never again be reborn into the dreaded cycle of suffering. In that moment, he became the Buddha—the awakened or enlightened one. Then, during the rest of his life he developed his beliefs,[12] and a community of devoted followers who spread his ideas wherever they journeyed.

The core doctrine of Buddha's atheism is *anatta*, which means no soul, no god, and no afterlife.[13] Whereas Hindus believed that Enlightenment

---

12. See Muller, *Dhammapada: The Essential Teaching of the Buddha.*

13. The discussion of Buddhism in this chapter applies to the Theravada branch that also is called Hinayana Buddhism or the small vehicle. We will not discuss the Mahayana (large vehicle) and Vajrayana (thunderbolt vehicle) forms that retain Buddha's core ideas and add other doctrines to his original philosophy, such as the three bodies of Buddha, the bodhisattva doctrine, or the concept of *sunyata* (extreme nonexistence).

led the soul (Atman) back to God (Brahman), Buddha rejected this view. Instead, he held that upon death an enlightened person enters a condition called emptiness, also expressed as nothingness. From then until now, the ultimate goal for Buddha's followers has been to escape from being reborn into suffering, which can be attained only through the spiritual experience of nirvana. Then, upon death, reincarnation ceases and the karmic energy that transmigrated from one life form to another disappears like the flame of a candle when it is blown out.

As Buddha developed the doctrine of anatta, he was asked how it was possible to reconcile the belief that there is no soul with the idea of reincarnation. If the soul does not exist, what is it that is reborn into successive life forms? In response, he added two additional doctrines. The first is called *anicca* and the second the *skandha*. Anicca means impermanence. Nothing lasts forever. Everything that exists depends on what came before it, which Buddha called dependent arising. Eventually, all material objects whether animate or inanimate perish. When a devout follower of Buddha's pathway experiences nirvana, the transmigrating skandha that is nothing more than a bundle of karmic energy disappears.

It is karma, which is defined as the law of moral conservation, that determines the form of the skandha's next reincarnation. A person's choices contribute directly to the accumulation of good or bad karma that drives the rebirth process. Craving for worldly success, envy, greed, and attachments to material possessions results in bad karma. In turn, this leads to perpetuating the cycle of reincarnation. The only way to end the cycle is through the attainment of nirvana.

Building on these basis concepts, Buddha developed his well-known Four Noble Truths and Eightfold Path. The four Noble Truths are that (1) all life is suffering, (2) suffering is caused by craving, (3) there is a way to end suffering by (4) following the Eightfold Path that consists of right beliefs, motives, speech, conduct, livelihood, effort, mindfulness, and meditation. For the devout follower of Buddha's way, there is but one aim: to achieve the mystical bliss of nirvana; and this is what defines Buddhism as a form of spiritual atheism.

Jainism

Jainism is another form of spiritual atheism even though the Jain worldview differs in major ways from that of Buddhism. The term Jainism is derived from the Sanskrit language, and it means conqueror. The founder of Jainism is Mahavira, which is a venerated title that means the Great Hero in the same way that the name Buddha means the Enlightened or Awakened One. Mahavira, whose real name was Nataputta Vardhamana, was born about the same time as Siddhartha the Buddha during the sixth century BCE. Like Siddhartha, as a young person in his thirties, Nataputta set out to attain nirvana that he eventually experienced later in his life. In so doing, his supporters called him the Great Hero who brought to an end, that is, conquered, the reincarnation cycle.

Like Buddha, Mahavira developed his unique atheistic views and coupled them with the pursuit of Enlightenment that he believed was the high point of the spiritual mountain that humanity had to climb to end rebirth. Also, like Buddha, he did not believe in a creator God who is immanent within the material universe. However, unlike Buddha, Mahavira believed in the existence of the soul called the *jiva* that is concealed within all of the material objects that comprise the cosmos.

At the same time, unlike Hinduism, he did not accept that each soul is identical in its essence or substance to all other souls. Rather, paralleling the monotheistic faiths described in the previous chapter, he believed that each soul is separate and unique. Furthermore, there are an infinite number of them, and they have always existed without beginning or end. He applied the same line of reasoning to the physical universe. Since God does not exist, there could not have been a transcendent power to create it or to be immanent within it. Thus, Mahavira concluded that just as all souls are eternal, so is the material universe.

Building on these basic assumptions about the nature of spirituality and materiality, Mahavira held a hierarchical view of the universe with materiality at the bottom and spirituality at the top. Materiality is evil, and spirituality is good. The spiritual goal is for each jiva or soul to rise from the material bottom to the spiritual top called *loka* that represents the spiritual apex of this hierarchical universe, where all the liberated jivas dwell together in perfect harmony, bliss, and knowledge. When a person has experienced nirvana in the Jain tradition, upon death the jiva floats to loka where it joins with other liberated jivas. See Figure 34.

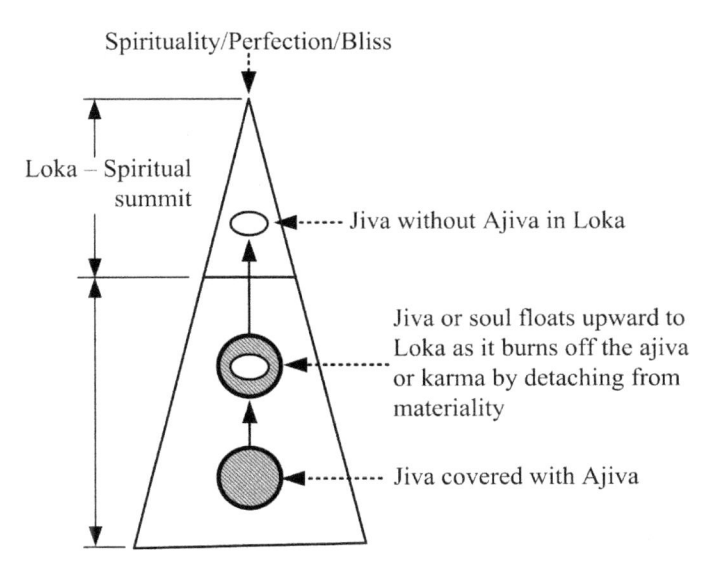

Spirituality/Perfection/Bliss

Loka – Spiritual summit

Jiva without Ajiva in Loka

Jiva or soul floats upward to Loka as it burns off the ajiva or karma by detaching from materiality

Jiva covered with Ajiva

Figure 34. Jiva Floats from Materiality to Loka.

The means of freeing the soul from its bondage to materiality involves pro-gressively burning away a sticky matter called *ajiva* that covers the soul or jiva. Ajiva is also defined as karma. Unlike Hinduism and Buddhism that include the concept of both good and bad karma, for Jains karma is always bad; and the Jain goal is to discard it through progressive detachment from materiality. As the karma that surrounds the jiva increasingly dissolves, the jiva floats higher and higher toward loka.

During his lifetime, Mahavira endured many self-imposed hardships as a result of his lifelong commitment to withdraw from all worldly pursuits and pleasures. In one final spiritual act of detachment, as an aging man he ended his life by refusing food and water and other basic necessities that would have kept him alive. For his followers, it was this last act of total detachment that sealed his exalted title as the Great Hero. Having achieved freedom from reincarnation, they rejoiced in the thought that his soul had finally arrived at the spiritual apex of the universe where it would dwell in peace and bliss with other liberated souls. Hence, spiritual atheism.

CONCLUSIONS

The non-monotheistic religions described throughout this chapter offered a greater diversity of beliefs than the monotheistic faiths we described in the previous chapter. They range from the pantheistic traditions of Hinduism, Taoism, and Sikhism that affirm that there is only one Ultimate Reality to those that assume the existence of multiple gods or spirits, including Hinduism and Shinto. As we have shown, atheism appears in two forms, secular and spiritual. Typically, secular atheism and agnosticism are associated with Western culture, whereas Buddhism and Jainism are the two major expressions of Eastern spiritual atheism.

In addition, the theological concepts of covenant and progressive revelation that play a central role in monotheism do not apply to the non-monotheistic faiths. Instead, knowledge, devotion to multiple deities or spirits, and mystical experiences of moksha or nirvana lie at the core of these traditions. The challenge that Christianity faces in our growing global village is how to relate to both the covenant and progressive revelation-oriented-monotheistic faiths that we discussed in the last chapter as well as the non-covenantal, non-monotheistic faiths we described in this one. Despite this theological and philosophical diversity, is there common ground that can serve to stimulate evolution toward greater interfaith dialogue and cooperation? We will address this question in the next chapter.

10

Christian Response to Religious Pluralism

INTRODUCTION

Table 11 summarizes the world religions that we have described in the previous two chapters. The table shows how they combine the four elements of materiality, spirituality, transcendence with or without intervention, and immanence to express their basic theological or philosophical beliefs.

Table 11. Combination of materiality, spirituality, transcendence, and immanence.

	Many Gods as Ultimate Spiritual Reality	One God as Ultimate Spiritual Reality
Religions where God (or Gods) is embedded within nature (Immanent)	Cell 1 Shinto Indigenous Religions	Cell 2 Hinduism (Monism and Dualism) Taoism Sikhism
Religions where God (or Gods) is outside nature (Transcendent)	Cell 3 Hinduism (Polytheism) Ancient Greek and Roman Polytheism (Henotheism) Indigenous Religions	Cell 4 Zoroastrianism Judaism Christianity Islam Baha'i

As cell 1 indicates, there are religions like Shinto and others existing in indigenous cultures where followers envision that many spiritual entities are embedded within nature or materiality. In Shinto they are called kami, and in native cultures they go by many different names.

Cell 2 includes Hinduism, Taoism, and Sikhism that are forms of pantheism. Hinduism incorporates both the monistic and dualistic types. All three of these three faiths hold that a single spiritual reality called Brahman, Tao, or Wehaguru, respectively, created materiality and is immanent within all the animate and inanimate objects that comprise the physical universe.

Cell 3 summarizes polytheism and henotheism, in which one of the many gods of the pantheon is dominant over all the others as in the case of ancient Greece and Rome. Like animism, polytheism and henotheism accept the existence of multiple deities with one critical difference. For polytheists, the deities are not hidden. They walk among us. They are often personified in the form of visual images, icons, and statues. Polytheists both past and present assume their gods possess extraordinary powers that can be used for good or evil and can influence fate and the course of human events. For the followers of animist traditions like Shinto, the many spirits are embedded within the physical objects of the universe.

Hindu polytheism shares much in common with the other polytheistic religions. However, for Hindus the many gods are only avatars of Brahman. The innumerable deities have no autonomous reality apart from being manifestations of Brahman. Each one stands as a separate pathway to the one Truth, which is that Brahman created the cosmos and is immanent within it in all of its various forms and functions.

Cell 4 includes the five monotheistic religions of Zoroastrianism, Judaism, Christianity, Islam, and Baha'i, which hold that there exists only one transcendent God who created the material universe but is not embedded within it. The Creator who goes by different names, such as Ahura Mazda, YHWH, Jehovah, Allah, or God, stands apart from the creation. All five of them include mediators (Zoroaster, Moses, Jesus, Muhammad, the Bab and Baha'u'llah), whom God chose specifically to reveal God's truth to humanity. Christianity is the only one of the five to consider its mediator Jesus Christ to be a savior who is equal to God in the Trinity. They also have their separate scriptures that record the unique historical events through which God intervened and made God's expectations known to humanity; and they view life as a testing ground that determines an individual's life-after-death destiny. Each person has only one life and not multiple lives through reincarnation.

We have not included Buddhism or Jainism in Table 11, because they exclude belief in a Transcendent Spiritual power. In chapter 9, we differentiated Eastern from Western forms of atheism. Buddhism and Jainism are associated with Eastern atheism because they set spiritual goals as the highest levels of human achievement. Western atheism does not. Also, agnostics who suspend judgment over whether or not there exists a Transcendent God who created the universe is related more typically to Western forms of atheism. In combination, the Eastern and Western forms of atheism and agnosticism along with the religions that are included in Table 11 comprise the vast majority of the world's theological and philosophical beliefs.

In the growing global village, this is the mosaic of worldviews that a spreading around the world as never before in human history. How devout followers of dissimilar traditions will react to each other is one of the major challenges in the foreseeable future. We will focus on three possibilities: exclusivism, pluralism, and inclusivism.

EXCLUSIVISM

The concept of religious exclusivism can be defined in simple terms. It means that among the many claims to Truth, only one of them is correct. Exclusivist thinking is binary, either or, and it is not confined to the followers of only one faith. In the history of the both Western and non-Western religions, there are countless examples of exclusivist positions. Limited space prohibits us from describing all of them here. Nonetheless, we will give examples from five of the world's most influential spiritual traditions starting with the three Abrahamic religions of Judaism, Christianity, and Islam and following with Hinduism and Buddhism.

Jewish exclusivist passages from the *Tanakh* or Hebrew Scripture also called the *Old Testament* contain references that define the Jews as God's chosen people.[1]

> Deuteronomy 7:6: For you are a people holy to the Lord your God; the Lord your God has chosen you out of all the peoples on earth to be his people, his treasured possession.

> Exodus 19:5: Now therefore, if you obey my voice and keep my covenant, you shall be my treasured possession out of all peoples.

1. Old and New Testament quotations taken from the New Revised Standard Version of the Holy Bible.

2 Samuel 7:23–24: Who is like your people, like Israel? Is there another nation on earth whose God went to redeem it as a people, and to make a name for himself, doing great and awesome things for them, by driving out before his people nations and their gods? And you established your people Israel for yourself to be your people forever.

In addition to the passages that designate that the Jews are God's chosen people, for Christians, the *New Testament* contains exclusivist verses that refer to Jesus as the only path to salvation.

John 14:6: Jesus said to him (Peter), I am the way, and the truth, and the life. No one comes to the Father except through me.

Acts 4:12: There is salvation in no one else (referring to Jesus), for there is no other name under heaven given among mortals by which we must be saved.

1 Timothy 2:5–6: For there is one God; there is also one mediator between God and humankind, Christ Jesus, himself human, who gave himself a ransom for all.

The third Abrahamic faith, Islam, also champions its own version of exclusivism that centers on Muhammad. Muslims consider him to be Allah's (God's) special Messenger, whom they call the Seal of the Prophets. Muhammad is the final prophet to whom Allah has revealed the perfect and definitive revelation as recorded in the following verses that are found in the Surahs or chapters of the Qur'an, Islam's sacred scripture.

Surah 9:73: O Prophet! Struggle hard against the disbelievers and the hypocrites, and be firm against them. Their home is Hell—Truly an evil place (to hide).

Surah 25:52: Therefore, do not obey the disbelievers, but work hard against them with the greatest strength, (with the Qur'an by your side).

Surah 48:29: Muhammad is the Messenger of Allah. And those who are with him are strong against the unbelievers, (but) compassionate within themselves.

Exclusivist claims are not limited only to the Abrahamic religions. When we shift to Eastern religions, we find them there as well. Exclusivism within Hinduism centers mainly on the caste system, which is described in the most important scriptures, the Vedas, the Laws of Manu, and the

Upanishads.[2] The four caste levels, also called the *varna* system, consists of the *Brahmins* (highest caste; scholars and priests), *Kshatriyas* (kings and warriors), *Vaishyas* (farmers, traders, and merchants), *Shudras* (laborers). A fifth group that falls outside of these four is called *Dalits* or Untouchables. They are outcastes who perform the most unskilled and menial tasks. Over time, numerous sub-castes emerged within this fourfold structure.

The long-established Hindu belief in reincarnation is closely tied to the caste system, which Hindus believe Brahman designed into the cosmos. Everyone is born into a specific caste level, and must perform exclusively the expected behaviors that are associated with that caste. The concept of karma is tied closely to the performance of the dharma or caste duties. If one performs the proper duties, then this leads to being reborn into a higher caste. The opposite holds true for the accumulation of bad karma that results in a lower rebirth including the possibility of being reborn as an animal or into the lowest group of untouchable Dalit outcastes also called Candalas. "Those who here are of delightful conduct will quickly attain a delightful womb—a Brahamana womb, a Kshatriya womb or a Vaisya womb. But those who here are of foul conduct will quickly attain a foul womb—a dog's womb, a pig's womb, or a Candala womb.[3] An individual cannot jump castes in order to speed up the reincarnation process that leads to moksha and the end of rebirth. Everyone transmigrates either up or down the caste system in an orderly sequence as prescribed in the Hindu sacred scriptures.

While Buddha rejected the Hindu belief in the Vedic caste system and the prescribed behaviors associated with each level, nonetheless, he set forth exclusivist claims of his own. These are tied to acceptance of his four Noble Truths and Eightfold Path as the exclusive pathway to nirvana. According to the Dhammapada, the most important scripture in the Theravada or original branch of Buddhism, called the Way of the Elders, anyone who takes refuge with Buddha, his teachings, and joins his movement "with clear understanding, sees the four holy truths: pain, the origin of pain, the destruction of pain, and the eightfold holy way that leads to the quieting of pain; that is the safe refuge, that is the best refuge; having gone to that refuge, a man is delivered from all pain."[4]

Following Buddha's way leads to nirvana that ends the suffering that he associated with reincarnation into biological or bodily existence. In his words, "Wise people, meditative, steady, always possessed of strong powers,

2. As social mobility in the urban areas of modern India increases, the historic ties to the caste system decreases. However, in the rural areas and villages, the traditional caste system remains strong.

3. Upanishads, pp. 163, 164 (v. 7).

4. Dhammapada, p. 46 (vv. 191–92).

attain to Nirvana, the highest happiness."[5] Or, "The sages who injure nobody, and who always control their body, they will go to the unchangeable place (Nirvana), where, if they have gone, they will suffer no more. Those who are ever watchful, who study day and night, and who strive after Nirvana, their passions will come to an end."[6]

After describing the exclusivist beliefs of these five religions (Judaism, Christianity, Islam, Hinduism, and Buddhism), we are left with several questions? Are the Jews God's only chosen people? Are others also chosen? Is the only one way to achieve salvation—either through Christ or by accepting the Qur'an as Allah's perfect revelation and Muhammad the Seal of the Prophets? Is it better to follow the Hindu caste system or Buddha's Four Noble Truths and Eightfold Path in order to end reincarnation? Are there other ways to stop the process of rebirth or to attain to the highest spiritual goal of moksha or nirvana?

In sum, for all exclusivists, only one view is correct. This means that by definition, the followers of all the other perspective are mistaken. Since all the faith communities of the world will remain in the minority for the foreseeable future, for a religious exclusivist, the vast majority of the world's population will continue to follow false beliefs.

PLURALISM

As we indicated earlier in this book, pluralism has a double meaning. In chapter 7, we defined pluralism in demographic and normative terms. Table 9 in chapter 7 lists the religions of the world descriptively by membership size and percentage of the total global population. The normative approach differs. It defines pluralism as peaceful coexistence. While definitions differ, at the most basic level, pluralism is a theory that there exist multiple definitions of Ultimate Reality. Or, pluralism involves "a situation in which people of different social classes, religions, races, etc., are together in a society, but continue to have their different traditions and interests."[7] Building on these two basic definitions, we can add: "Religious pluralism is basically accepting that all religions are equal, valid, and ultimately lead to God. It is the coexistence of various religions under the same roof, and celebrating the presence of other religions without losing one's own identity."[8] These definitions

5. Dhammapada, p. 11 (v. 23).

6. Dhammapada, pp. 54–55 (vv. 225–26).

7. *Merriam-Webster.com*, s.v. "pluralism."

8. "Understanding the Concept of Religious Pluralism with Example," SpiritualRay, https://spiritualray.com/understanding-concept-of-religious-pluralism-with-examples.

combine both the descriptive and normative dimensions of the concept of religious pluralism as we are using it in this chapter.

In the above section on Exclusivism, we quoted from the scriptures of five of the world's best-known sacred traditions. Can we find in these same scriptures passages that support a pluralistic view of religion? The answer is yes, as we show below starting with Judaism. The Jewish pluralist position is based on the flood narrative that appears in Genesis where the forty-day deluge that covered the surface of the earth killed every land creature except Noah, his family, and the animals that boarded the ark with them. According to this perspective, all modern humans are descendants of Noah in whose name God created a covenant that includes all of humanity. Here are the relevant passages:

> Genesis 9:1: God blessed Noah and his sons, and said to them, "be fruitful and multiply, and fill the earth."

> Genesis 9:8–9: Then God said to Noah and to his sons with him, "As for me, I am establishing my covenant with you and your descendants after you, and with every living creature that is with you."

> Genesis 9:11,12: "I establish my covenant with you, that never again shall there be a flood to destroy the earth." God said, "This is the sign of the covenant that I make between me and you and every living creature that is with you, for all future generations: I have set my bow (rainbow) in the clouds, and it shall be a sign of the covenant that is between me and you and every living creature of all flesh."

In addition to the specific 613 laws that are enumerated in the Torah and pertain to the Jews as God's chosen people, the general Noahide covenant laws that appear in chapter 9 apply to the whole of humanity regardless of religion, race, nationality, etc. These include acknowledging that there exists only one God who shall not be cursed along with prohibitions against murder, stealing, eating living animals, and abusing human sexuality. The Noahide covenant also calls for establishing courts of law that promote justice for everyone throughout the world regardless of a person's national or cultural origin or social background.

In addition to the Noahide references in Genesis, which support the pluralist position within Judaism, there are parallel New Testament passages that do the same for Christianity.

Matthew 7:21: Not everyone who says to me, "Lord, Lord," will enter the kingdom of Heaven, but only the one who does the will of my Father in heaven.[9]

John 13:34–35: I (Jesus) give you a new commandment, that you love one another. Just as I have loved you, you also should love one another. By this everyone will know that you are my disciples, if you have love for one another.

Matthew 22:34–40: When the Pharisees heard that he (Jesus) had silenced the Sadducees, they gathered together, and one of them, a lawyer, asked him a question to test him. "Teacher, which commandment in the law is the greatest?" He said to him, "You shall love the Lord your God with all your heart, and with all your soul, and with all your mind. This is the greatest and first commandment. And the second is like it: You shall love your neighbor as yourself. On these two commandments hang all the law and the prophets."[10]

Romans 13:8, 9: Owe no one anything, except to love one another; for the one who loves another has fulfilled the law. The commandments, "You shall not commit adultery; You shall not murder; You shall not steal; You shall not covet; and any other commandment, are summed up in this word, Love your neighbor as yourself."

1 Corinthians 13:13: And now faith, hope, and love abide, these three; and the greatest of these is love.

1 John 4:7–8: Beloved, let us love one another, because love is from God; everyone who loves is born of God and knows God. Whoever does not love does not know God, for God is love.

In combination, these New Testament verses broaden the narrow passages that we cited in the above section on Exclusivism. They point to a love ethic that includes the whole of humanity regardless of personal or collective differences. Jesus reference to loving God fully and one's neighbor as oneself along with Paul's and John's emphasis on love as a new commandment that

9. In Matt 25:31–46, Jesus identifies doing God's will with specific actions. These include feeding the hungry, giving drink to the thirsty, providing clothing and shelter for the needy, welcoming the stranger, taking care of the sick, and visiting those held in prison.

10. This passage is repeated in Mark 12:28–34 and Luke 10:25–28.

is even greater than faith and hope points toward not giving preference to some groups over others. Furthermore, Jesus's parable of the Good Samaritan (Luke 10:29–37) extends caring for others beyond one's narrow group boundaries.

Islam's sacred scripture, the Qur'an, also contains surahs that parallel the pluralist passages found in the Old and New Testaments.

> Surah 5:48: If Allah had so willed, He would have made you (all) a single People, but (His plan is) to test you in what He has given you: So work hard as if (you are) in a race in all good deeds.

> Surah 16:93: And if Allah had so wanted, He could make you all as one People (one Nation).

> Surah 2.256: Let there be no force (or compulsion) in religion: Surely—Truth stands out clear from error: Whoever rejects evil and believes in Allah has held the mot trustworthy hand-hold that never breaks. And Allah is all Hearing, All Knowing.

> Surah 5:32: If anyone killed a person . . . it would be as if he killed all mankind (the people): and if anyone saved a life, it would be as if he saved a life of all mankind (the people).

> Surah 34:4: That he may reward those who believe and work deeds of righteousness: For such (persons there) is forgiveness and most generous ways (and means) to live.

> Surah 109:6: To you be your religion, and to me (be) my religion.

This combination of Surah verses shares much in common with those from the New Testament, and in some cases makes the case for pluralism even stronger. On more than one occasion, Muhammad recognized that God (Allah) could have made everyone as one people or nation if God had chosen to do so and that there is no compulsion in religion. He recognized that others had their religion and he had his and that killing or saving even one person was equivalent to killing or saving all people. He acknowledged God's generosity and forgiving nature and counseled his followers to compete with each other in doing acts of goodness and justice.

Thus, while the sacred scriptures of all three of the Abrahamic religions (Old Testament, New Testament, and Qur'an) contain numerous passages that support exclusivism, they also contain other verses that promote religious pluralism. Can we say the same for Hinduism and Buddhism?

In support of Hindu pluralism, there are alternatives to the exclusivism that requires reincarnating through the traditional four stages of the caste system in a prescribed manner from the bottom up—one level at a time—or better stated, only one level per lifetime. At its core, the Hindu approach to religion is innately pluralistic, because it champions the belief that there are many pathways to the one Truth that is Brahman.

One of the most well-known and colorful images through which Hinduism portrays its commitment to pluralism is the parable of the blind men and the elephant that originated in ancient South Asia and became diffused throughout the three main religions of Hinduism, Buddhism, and Jainism.[11] In the parable, a group of five blind men come upon an elephant that they have never before encountered. Each blind man touches a different part of the elephant and has a limited point of view. The moral of the story is clear. Each sees the truth of the elephant from a limited perspective. The elephant (total Truth) is greater than any single experience (limited truth) of only a part of it.

Furthermore, as we discussed in chapter 8, there are four main paths that lead to the spiritual goal of Enlightenment that results in returning the Atman or soul back to Brahman or God. In addition to following caste and life cycle requirements, the other three involve using the intellect to acquire sacred knowledge, meditating, and showing religious devotion. Thus, just as the three Abrahamic religions of Judaism, Christianity, and Islam contain both exclusivist and pluralist options, the same can be said of Hinduism and, as we will show next, Buddhism.

There are two main branches of Buddhism.[12] The first is called Theravada, which means the way of the elders. It is also called Hinayana or small vehicle. After Buddha's death, his followers fanned out from South Asia to Central and East Asia; and eventually they developed in new branch known as Mahayana—or large vehicle. Central to the Mahayana teaching is the doctrine of the Bodhisattva, which means a Buddha in the making. Any individual can become a bodhisattva who chooses to postpone a final nirvana by transmigrating through countless reincarnations until all other beings achieve their own enlightenment.

---

11. Strong, *Udana*. One of the earliest written versions of the elephant parable can be traced to the Buddhist text *Udana* 6.4, dated to around the middle of the first millennium BCE, but originally it most likely was transmitted orally. Other metaphors such as a tree with many leaves and a house with many rooms are also used to convey images of religious pluralism.

12. Some scholars identify a third branch called Vajrayana that is identified with Tibetan Buddhism, whose best-known spiritual leader is the Dalai Lama. We include Vajrayana as part of the Mahayana branch.

The Great Vow of compassion, also known as the Bodhisattva Vow, captures the essence of this doctrine.

> I take upon myself the burden of all suffering; I am resolved to do so; I will endure it. I do not turn or run away, do not tremble, am not terrified, nor afraid, do not turn back or despond. . . . All beings I must set free. . . . I must rescue all these beings from the stream of Samsara, which is so difficult to cross, I must pull them back from the great precipice, I must free them from all calamities, I must ferry them across the stream of Samsara. I myself must grapple with the whole mass of suffering of all beings.[13]

The Bodhisattva Vow builds on Buddha's First Noble Truth that all life is suffering and elevates the norm of compassion to the top of the list of desirable character traits. The Mahayana or great vehicle school builds on Buddha's basic beliefs and adds others to them. Not only is the original Buddha and his teachings there to help one cross the stream of Samsara or suffering (the small vehicle) but other Buddhas or bodhisattvas are there to help as well (the large vehicle). The bodhisattva's goal is to reach out to anyone through repeated rebirths regardless of their religion, race, nationality, or status in order to show them the pathway that leads to enlightenment and the cessation of suffering through reincarnation.

INCLUSIVISM

In addition to exclusivism and pluralism as two distinct ways in which the followers of Judaism, Christianity, Islam, Hinduism, and Buddhism, as well as other world religions can interact with each other in the expanding global village there is a third alternative: inclusivism.

Of the three ways in which member of the world's diverse faiths might relate to each other, inclusivism is the least clear-cut. This is because inclusivism can be interpreted in ways that sound like pluralism, as we show below; and it is often difficult to distinguish between them. Exclusivism differs from inclusivism by asserting that only one of the world's diverse religion is true and all the others are untrue. Neither pluralism nor inclusivism is this restrictive.

The main challenge that the inclusivists face is to carve out a distinct position. How might this be done? Unlike exclusivists, who claim that only one perspective is true or pluralists who assert all are equally true, one form of the inclusivist method involves creating a sliding scale of relative degrees

---

13. Young, *World's Religions*, 150.

of truthfulness. The religion that possesses the most or total Truth goes at the top of the list, while the others are positioned hierarchically with the one with the least amount of truthfulness at the bottom.

In addition to the sliding scale method, there are two other ways to approach inclusivism. The first involves identifying the strengths of existing religious groups and synthesizing them into a comprehensive faith that replaces the others. A second entails creating an entirely new religion that supersedes all existing religions. Given that religious pluralism is likely to continue, these two alternatives hold little promise for the future. Furthermore, any new religion or synthesis of beliefs from existing religions is likely to become just one more religion among the many, as Sikhism did when combining Hindu and Islamic elements. Thus, it would appear that among the inclusivist options, the sliding scale approach has the most potential.

However, there are major challenges with this method given the diversity of faiths that populate the planet. Any attempt to rank order the world religions according to degrees of truthfulness invites the question: Which religion belongs at the top of the list and why? Since there is no universal or objective theological or philosophical standard for comparing different religious beliefs, the choice of placing one religion above the others will always be arbitrary. Do monotheistic beliefs take preference over the others? If so, which one comes first? Christianity? Islam? Other? If monotheism does not belong at the top, which alternative does? Pantheism? Polytheism-henotheism? Animism? Spiritual atheism? Non-spiritual atheism or agnosticism? Hinduism over Buddhism? Taoism over Shinto? And so on.

Despite this rank ordering challenge, inclusivists have searched for ways to measure the relative degrees of truthfulness as a middle position between exclusivists and pluralists. At the same time, they have not endeavored to arrange all of the world's religions on a sliding scale of truthfulness—and for good reason. Given the vast number of existing world religions, such an undertaking would seem impossible. Nor does the existence of multiple world scriptures offer much guidance, because they were written in different languages and in diverse cultures with limited interaction prior to development of global communication and mass transportation technology.

In the process of searching for an alternative between exclusivism and pluralism, inclusivists have developed an approach that combines their own faith commitments with those of others while at the same time avoiding the impossible task of developing a comprehensive list that includes all the religions. For example, the Catholic theologian Karl Rahner has advanced a position called the "anonymous Christian." In his view, Christ is the ultimate spiritual standard for measuring the experiences and traditions of non-Christian religions. Christ fulfills the divine mystery or Ultimate

Reality toward which all of the world religions strive. God is also at work in the other religions, and they may serve as alternative pathways to salvation. If the believers of other faiths practice their ideals in good conscience, they may receive God's grace. Rahner refers to the faithful followers of other religions as anonymous Christians.[14]

As a result, it is not necessary to provide an exhaustive hierarchical list that ranks the religions of the world according to their degrees of truthfulness. While Rahner believes that Christianity heads the list of the world faiths, he also affirms that God desires and grants salvation to people in different religions. Thus, he avoids exclusivism because he accepts that God may have visited grace on others before Christianity emerged as well as on non-Christian people after the church began; but he is not a pluralist because he does not define all religions as equal.

Upon closer look, Rahner's inclusivism could be interpreted as a subtle or soft pluralism, and here is why. If the followers of both Christian and non-Christian faiths could be recipients of God's grace, does this not make all of them equal? Rahner takes seriously that the realm of religion is comprised of a plurality of faith communities and that God is active in all of them and not just only one. As a Christian theologian, based on the Trinity, he exalts Christ as the perfect embodiment of the divine presence in humanity and the standard for interfaith comparisons. His personal commitment to Christ does not lead him to the conclusion that being a non-Christian means that the devout followers of other faiths are excluded from God's gift of grace. Instead, he calls them anonymous Christians who are also capable of receiving God's salvation.

The inclusivist approach of the anonymous believer is not unique to Rahner. It has parallels among non-Catholic Christians as well as devout followers of non-Christian religions. While space does not allow for us to discuss them here in detail, we can identify the pattern of thinking that appears in all of them: Religion X (Christianity, Hinduism, Buddhism, etc.) ranks at the highest level, and the other religions are anonymous subsets of X. This form of inclusivism can also be characterized as having a foot in two camps with one foot being bigger than the other. The first camp refers to the religion of one's choice (the bigger foot), the second to other religions (the smaller foot).

This approach to inclusivism is a compromise. It takes religious pluralism seriously without asserting that only one religion is true or all are equal. This is its main strength. However, it also has a major weakness that relates to its core concept of anonymity. If the proponents of any given religion can

14. Rahner, *Christianity and Other Religions*.

define the faithful followers of other religions as anonymous believers of their religion even though they might be unaware of it, then the members of other religions can do the same in reverse. Hindus can assert that faithful non-Hindus are anonymous Hindus. Muslims can say that non-Muslims are anonymous Muslims, and so on for other religions.

Furthermore, claiming that the followers of other religions are anonymous believers of one's own religion in effect relegates other religions to a secondary status. While this might not be the intent, it is the result. It also leads to an indirect invalidation of other religion's spiritual experiences by viewing them through the doctrinal lenses of different religions. Nirvana could be seen as an anonymous, mystical experience of Christ, Allah, Brahman, or other. Moksha could be viewed as an encounter with the Zoroastrian God Ahura Mazda or the Tao. And so on. Thus, in the search to explain how God or some other understanding of Ultimate Reality is present in all the world religions, the anonymity approach to inclusivism ends up being dualistic because it holds that only one religion fulfills God's purposes fully (level 1) and that the faithful followers of the other religions could be anonymous believers of that religion even though they do not self-consciously identify with it or are unaware of it (level 2).

Baha'i offers an alternative approach to inclusivism but also contains dualistic elements in its two-cycles view of history. As indicated earlier, the first cycle, called the Adamic cycle, contains numerous dispensations (time periods) and special manifestations (messengers) from Adam through Muhammad through whom God revealed new spiritual teachings. Baha'is consider them all equal and cumulative from start to finish.

The second cycle, the Cycle of Fulfillment, started with the revelations of Bab and Baha'u'llah. It builds on the combined revelations of the first cycle and will end when universal peace and justice are established on earth. The second cycle takes God's progressive revelations to new levels without displacing earlier ones. In effect, Baha'i is a combination of pluralism (all revelations are equal during the Adamic Cycle) with inclusivism (the Cycle of Fulfillment is a newer time period that will culminate in world peace).

Rahner's anonymous believer position and Baha'i's two-cycles view of progressive revelation represent inclusivist alternatives to both exclusivism and pluralism. In contrast to believing that only one religion is true or that all are equally true, Rahner's and Baha'i's two-level approaches start with specific, although different, religious assumptions. They explain their own spiritual traditions through those assumptions and then use them as a theological filter or framework for interpreting how God is present in the faith experiences of others.

One of the major critics of the inclusivist approach to ranking the world religions through the theological assumptions of any one religion is the well-known philosophical theologian John Hick, who holds that all such approaches are biased in favor of a dominant religion even though they genuinely strive to make room for other religions as well.[15] In his view, the only viable option is pluralism that starts with the assumption that all religions are equally true. Given the wide range of viewpoints that include monotheism, pantheism, polytheism/henotheism, animism, atheism (spiritual and non-spiritual), and agnosticism, we are led to ask: how is this possible?

Here is Hick's answer. He starts by presuming that there is one Ultimate Reality that is present in all the world religions despite their diversity. Beyond the many lies the One. He does not use the word God to define this Ultimate because religions differ in their definitions of God (Hinduism, Christianity, and Islam), and some spiritual atheists' traditions do not include belief in God at all (Buddhism and Jainism). Since use or non-use of the word God varies so widely, there is need for an alternative name for the Ultimate, which Hick calls "the Real." In his understanding, the Real is the everlasting and inexhaustible Ultimate Reality that both transcends and is immanent within all the world's religions despite their differences.

From this perspective, Hick's description of the Real parallels the elephant parable. Just as each blind person touches only a part of the elephant, every religion is only one of many theological or philosophical interpretations of the Real. The religions of the world differ because they emerged in different historical and cultural contexts that include dissimilar languages that followers use to describe them. This means that no religion's view of the Real is absolute or exclusively true or necessarily more complete than others. No single religion encompasses the totality of human encounters with the Real. All of them are finite expressions of an infinite and inexhaustible source, the Real, that is greater than all of them in the same way that the elephant exceeds the blind persons' descriptions of some of its parts.

Does this make Hick a relativist? In his interpretation, the answer is no. Relativism is the view that there is no external standard or common framework of comparison that transcends the various truth claims of each world religion. For relativists, universal norms or concepts do not exist, and the authority of each religion is confined to the cultural setting in which it emerged and is practiced. If Hick is not a relativist, then what is he? The answer is that he is a pluralist who believes that the Real is present in all religions even though they differ in their perceptions and descriptions of it.

---

15. Hick, *Interpretation of Religion*, 236–46.

In effect, this is the common assumption that is shared by all pluralists however they define the relationship of the finite to the infinite. The many smaller truths are always seen as manifestations of the one larger Truth that transcends and/or is immanent within all of them. This is their "theological leap," so to speak, even though there is no scientific evidence either to verify or falsify the existence of such an Ultimate Reality regardless of what it is called. As a result, Hick along with pluralists in general hold that pluralism is the preferred alternative to claiming that only one religion is true or that they can be ranked according to degrees of truthfulness.

CHRISTIANITY AND RELIGIOUS PLURALISM

No matter how we define exclusivism, pluralism, and inclusivism, in the final analysis it boils down to explaining the relationship of the One to the many. Exclusivism holds that only one of the many is true; pluralism assumes that the many are equally true; and inclusivism regards one of the many as the truest and interprets other religions through it. Where do Christians stand? The answer is: There are Christian exclusivists, Christian pluralists, and Christian inclusivists. Just as there are many world religions, there are many Christian responses to the question of how Christianity should relate to other religions as religious pluralism continues to spread around the globe.

In the final analysis, all Christians will respond one way or another depending on their view of God and Christ. Exclusivists hold that only those who accept Christ as their personal savior will be rewarded in the afterlife with heaven. As we have shown, there are biblical passages that support this position. In the pluralist view, anyone who shows love toward others knows God, and they too will be rewarded by God. There are also biblical passages to uphold this perception. For inclusivists, it is a major challenge to find biblical verses in support of this option, because global religious pluralism as we are experiencing it currently did not exist in the ancient world when the Bible was written. As a result, theologians like Rahner have created innovative theological concepts like the anonymous Christian to explain the relationship of Christianity to other world faiths and to support the belief that God's gracious gift of salvation is available to non-Christians as well as Christians.

If we cannot determine that there is one, and only one, correct Christian position, can we rank all three of them on a scale of most to least Christian? The answer is: probably not—for this reason. The adherents of each position base their views on different sources of authority, and this is a highly subjective process related to many family, sociological, psychological, and other factors that limited space does not permit us to discuss here.[16] Furthermore, there is no objective standard for selecting which sources of authority are the most authoritative or more authoritative than other sources, including selecting specific biblical passages to make the case for a particular position. As we have shown, exclusivists quote as authoritative the passages that claim that the only way to the Father is through the Son: pluralists point to the broader love passages; and inclusivists turn to modern theologians like Rahner.

In addition, defending a particular perception with remarks such as "The Bible says . . ." or "The church says . . ." does not resolve the authority issue, because different people can quote different passages, doctrines, or traditions to justify dissimilar positions. Also, asking the question "What would Jesus think or do?" does not remove this difficulty, because different people think differently about what Jesus might think or do; and they would pick out different authority sources and biblical references to support their interpretation of the mind of Jesus. Thus, it appears as though Christians differ not only in their views of how Christianity relates to non-Christian religions but how different Christian views relate to each other. Thus, for Christians, the dilemma is both inter- and intra-religious, which is, of course, not unique to Christianity. It applies to the other world religions as well.

Where does this leave us with regard to the world diverse spiritual communities in general and Christianity in particular on the modern world challenge of how religious pluralism might evolve in the direction of greater interfaith harmony? We will answer this question in chapter 12. However, before doing so, we will turn to chapter 11, where we will address the issue of the relationship of Christianity to the four big questions that we listed in the first chapter. This will set the stage for our final chapter, chapter 12, in which we describe how the evolution of faith is related to Christianity, science, and the world religions. In the process, we establish a new and fifth alternative, called Evolutionary Pluralism, that we compare to the four positions of Young Earth Creationism, Old Earth Creationism, Intelligent Design Creationism, and Evolutionary Creationism that we described in chapters 5 and 6. We are now ready for chapter 11.

16. Many books and articles have been writing of the sociological and psychological factors that contribute to how and why individuals develop their religious beliefs. No doubt, the family plays a formative role in this process, as do the diverse cultural standards and patterns that vary from society to society. Our main focus in this chapter is on the three main ways in which Christians can relate to non-Christians and the sources they choose to justify their positions.

11

Christianity and Four Big Questions

INTRODUCTION

In this chapter, our main focus will be to integrate the four big questions with our earlier discussion of the relationship between Christianity and science and Christianity and religious pluralism. As we listed them in chapter 1, these four questions involve (1) where the universe came from and how it operates, (2) whether there exists a spiritual power that is greater than the universe, and—if so—how it relates to (3) the universe and to (4) humanity. After addressing the first of the four, we will follow up sequentially with the other three and show how they are all interconnected.

WHERE THE UNIVERSE CAME FROM
AND HOW IT OPERATES

In our response to the first big question of where the universe came from and how it operates, we have shown that science has evolved from the premodern to the present modern stage. Our goal in this chapter is not to reiterate all those details but to draw additional insights from what we have learned. In this section, we describe three such insights. First, as a result of the astonishing discoveries that numerous scientists made during the past four centuries, starting with Copernicus's sun-centered theory and Galileo's telescopic observations of the moons of Jupiter, we know now that in terms of sheer age and size, our universe is so immense that our limited human imaginations find it nearly impossible to grasp.

As the bell-shaped surface in Figure 35 shows, the big bang occurred 13.8 billion years ago, and the approximate diameter is 92 billion light years. The time of 13.8 billion years is way beyond our comprehension. To see why this is so, we can scale down 13.8 billion years to just one year. Assuming the average human lifetime is eighty years from birth to death, the eighty years in relationship to 13.8 billion years is compressed to just 0.18 seconds on the scaled time of one year.[1] For all intents and purposes, the age of the universe is not conceivable in terms of our day-to-day experiences or even in terms of one human lifetime.

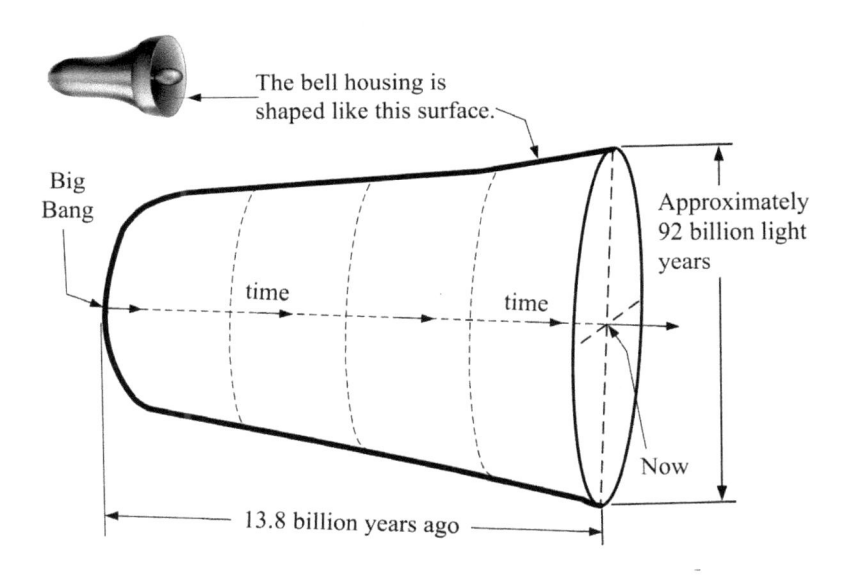

Figure 35. Summary of the size and history of the universe.

As Figure 35 also shows, the approximate size of the universe is 92 billion light years across. If a beam of light could be shaped to travel around the earth's equator, it would take a little less than one second for the light

1. Eighty years compared to 13.8 billion years is $80/13.8 \text{X} 10^9 = 5.80 \text{X} 10^{-9}$, using scientific notation for the numbers. On the one-year scale there are $3.15 \text{X} 10^7$ seconds. $(5.80 \text{X} 10^{-9}) \text{ X } (3.15 \text{X} 10^7) = 0.18$ seconds. So, eighty years on the first scale compresses to 0.18 seconds on the 1-year scale.

to circle the equator seven times; and this is equal to one light second. Furthermore, at the speed of light (186,000 miles per second), it takes 5.5 hours for the sun's light to travel to Pluto, which is at the edge of our solar system.[2] This means that at the speed of light, it would take about eight hundred round trips for the sun's light to travel to Pluto and back in one year.

Thus, like the age of the universe at 13.8 billion years, the distance of one light year is impossible for humans to comprehend. When we shift from a single incomprehensible light year to 92 billion light years, which is the approximate width of the universe, in human terms, the enormity of the universe is completely inconceivable. Furthermore, the width of the universe is enlarging. Modern science has detected that the hundreds of billions of galaxies that populate the universe are moving away from each other, resulting in the continuing expansion of the cosmos. Additionally, the rate of expansion is accelerating. Given what we know about gravity as the only additive force field at these unimaginably large scales, this makes no sense. Scientists explain this acceleration in terms of a "dark energy," although no one really knows what this is. At this point in time, dark energy is still very much an unresolved mystery.

In addition to the enormous age and size of the universe at the macro level, the subatomic micro world is equally complex. To use one example, particles such as electrons, quarks, and Higgs bosons sometimes behave as if they were distributed in space as waves and at other times as if they were localized clumps of matter (particles). When a pitcher throws a baseball (particle) toward home plate at ninety miles per hour and the batter swings and makes contact, that baseball does not suddenly act like a wave, causing a spray of baseball mist around home plate. Alternatively, when a wave on a lake intersects a pier, the wave does not act as if it were a baseball smashing into the pier while the surrounding water is undisturbed. The wave is spread out and not like a baseball.

A second key insight that emerges out of our discussion of how the universe began and operates is that the process of scientific discovery is neither linear nor predictable. For example, for thousands of years during the premodern science era, with the unaided eye no one could see the moons that circled Jupiter. When Galileo observed them for the first time through his newly built telescope, he opened the door to entirely novel images of the cosmos. His findings were logically inconsistent, that is, counterintuitive, to the geocentric model in which all heavenly bodies were believed to move around the earth during the premodern science era. Instead, he discovered

2. The edge of the solar system is assumed to extend out to the noncircular orbit of Pluto or about 3.67 billion miles.

something that was genuinely new and unexpected;[3] and this led eventually to replacing the Ptolemaic geocentric theory with the sun-centered, heliocentric model of our solar system.

Following in Galileo's footsteps, a succession of remarkable scientists that included Kepler, Newton, Einstein, Heisenberg, Hubble, and Darwin added new pieces to the puzzle of how our cosmos began and operates. In effect, the history of science from the premodern to the modern era is the story of solving many of the mysteries of the material universe. As this process unfolded, most of the major breakthrough observations ran contrary to scientific perspectives that existed at the time.

The reason Galileo exemplifies this process is that no one foresaw that his novel telescopic observations would revolutionize our understanding of the universe. His discoveries were not consistent with the prevailing geocentric theory. Rather, they were counterintuitive. Following Galileo, Kepler discovered that observed planetary motions were better described by elliptical, not circular, orbits. This was counterintuitive at that time, because all heavenly bodies above the earth were understood to move along circular paths. Circles were thought to be perfect, uncorrupted, and uncontaminated shapes, consistent with the essence of those heavenly spaces.

In quantum physics, it came to be understood that a particle's position and momentum (associated with its speed) cannot be simultaneously known with complete precision. This is known as the Heisenberg uncertainty principle. This was very counterintuitive to the prevailing Newtonian view that we can simultaneously know a particle's position and momentum with arbitrary accuracy. All we had to do was devise, design, and build better instrumentation and measure with greater resolution. However, according to the Heisenberg uncertainty principle, precise measurement cannot (even in principle) go beyond a specific stated limit, no matter how good the instrumentation and how careful the measurements are made.

Einstein provides us with a supreme example of another counterintuitive discovery. Prior to positing a theory of general relativity, Newton's universal law of gravitation imagined gravity as a force that instantly extends throughout the entire universe from each planet, star, and galaxy, and, indeed, every particle. Einstein's unexpected new idea is that time (one dimension) and space (three dimensions of length, depth, and width) combine into four-dimensional space-time. A warpage occurs in space-time depending on the amount of nearby mass and/or energy. The greater the mass (everything else constant), the greater the warpage and vice versa for

---

3. This is the heliocentric understanding (the earth moves around the sun) that Copernicus proposed earlier in 1543 CE.

less mass (and/or energy). General relativity has passed every single experimental and observational test since Einstein first proposed this outstanding achievement in 1915. As we have shown, science abounds with these kinds of surprising and counterintuitive findings.

Our third insight is as important as the first two, and it cannot be overstated: science is self-correcting. As we have shown in chapter 6, the National Academy of Sciences is the United States' premier organization that has defined the criteria and method for the conduct of modern scientific research. NAS guidelines start with an inductive approach to the study of the material universe through observations and the accumulation of evidence. They also include the need for repeated testing and experiments to confirm, modify, or falsify scientific hypotheses regarding natural phenomena, the quantification of findings, peer review, predictability, and so on. The link between the NAS and answers to the first of the four big questions is direct and visible in terms of our current understanding of the age and size of our universe.

At this point we can connect the counterintuitive process of scientific discovery with the necessity of following the methods for conducting scientific research. We use as our example the shift that occurred from the geocentric to the heliocentric view of the universe. In 1543, when Copernicus first proposed the sun-centered theory, he was well aware that this conflicted with the centuries-old earth-centered perspective. After Galileo's unexpected telescopic observations, a new perception of the cosmos began to emerge, which competed with the existing Ptolemaic view. As we have shown, similar patterns appear in the cases of Kepler, Einstein, Heisenberg, Hubble, and Darwin.

When competing scientific hypotheses exist, our natural curiosity leads us to ask: Which view is correct? The only way to answer this question is through verifiable observations and experiments that can be conducted over time by following scientific methods. The findings of Galileo were counterintuitive to the intuitive or taken-for-granted view of the universe that existed at that time. With hindsight, we know now that Copernicus was correct, and Ptolemy was not. This outcome would not have occurred apart from the self-correcting nature of science itself.

As we have shown repeatedly in chapters 2–6, it is precisely by following guidelines for scientific research that scientists have confirmed or falsified a given hypothesis regarding how the laws of nature operate in any given field of inquiry. When there exist competing hypotheses, as in the case of the geocentric versus the heliocentric theory, along with others that followed, confirmation of one or the other with modifications if necessary or falsification of both can be determined only through repeated observations, experimentations, predictions, peer review, and so on. Another way to say this is that if it were not for science's self-correcting nature, we would never have improved our understanding of how the universe began and operates (the first big question).

We are now ready for the second big question.

SPIRITUAL POWER GREATER THAN THE UNIVERSE

The second big question on the existence of a higher spiritual power that is greater than the material universe is related directly to the first big question about where the universe came from and how it operates. At first glance, it might seem as though these two areas are not connected to each other. However, the opposite is true, as we will show. Once again, we will not re-iterate all the details that we included in chapters 7–10 but identify new insights that result from the combined descriptions of the various world religions.

The first insight related to the second big question of a spiritual power greater than the material universe is tied directly to modern science's discoveries about how our universe came into existence and operates. As we have shown in the above section, the discoveries of modern science reveal to us that our universe is massive—13.8 billion years old; and large—92 billion light years wide and expanding. In the face of such a vast and awesome cosmos, our curiosity leads us inevitably to wonder: assuming that there exists a spiritual power greater than the universe, what must this power be like? The answer: even more amazing.

Thus, we can combine our responses to the first two big questions in the following way: (1) if we accept that there exists a spiritual power that goes beyond the material universe, and (2) if the universe is as enormous as modern science reveals it to be, (3) then it follows that the creative power that brought the universe into existence is greater than the universe itself. Once we accept the existence of such a spiritual power, we move to the next question: How are we to understand this power? Our search for an answer to this question shifts us away from the sphere of science and into the realm of religion.

Figure 36 summarizes the diversity of world religions that we discussed in chapters 7–10. As Figure 36 shows, we have grouped the world's religions according to their collective acceptance of the universal reciprocity norm and how they have responded to the search for the Ultimate. We consider that both of these elements are common to all religions. The theistic religions fall under either monotheism or non-monotheism. We have included Zoroastrianism; the three Abrahamic faiths of Judaism, Christianity, and Islam; and Baha'i under that heading of monotheism. Under non-monotheism, we have listed Hinduism, Taoism, and the traditional Chinese spirituality, Sikhism, and Shinto. The Eastern religions of Buddhism and Jainism link to spiritual atheism, and the Western forms of non-spiritual materialism include atheism and agnosticism.

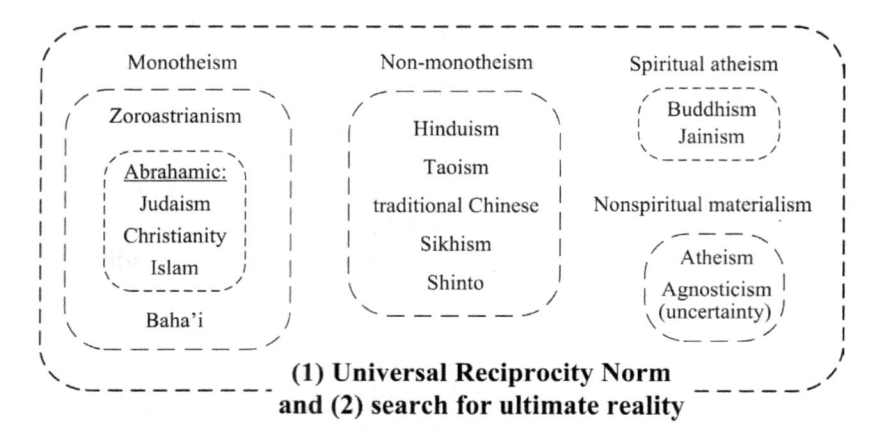

Figure 36. Summary of world religious diversity.

Our second insight extends directly from Figure 36, which outlines the broad diversity of beliefs that exist among the globe's many religions in their perceptions of an ultimate power that is greater than the physical world. Another way to say this is that there is no common understanding among the followers of the world faiths about the nature of the Ultimate. Unlike the large majority of current scientists who accept the modern scientific view of the immense size and age of the universe, no parallel consensus exists among the followers of the world religions with regard to an Ultimate spiritual force. Our descriptions of the diverse faiths that populate our planet include multiple forms of monotheism, pantheism, polytheism and henotheism, animism, and spiritual and non-spiritual atheism.

Why should this be so? There are two main reasons. First, the world religions that differ in their perceptions of the Ultimate emerged thousands of years ago under conditions of social and geographical isolation, such as Zoroastrian monotheism and Taoist pantheism. Others that share common concepts such as the Abrahamic faiths of Judaism, Christianity, and Islam and Baha'i arose consecutively within a common cultural setting—in this case, the Middle East. Currently, as a result of the worldwide web of electronic communication and mass transportation, the trend toward greater interaction between the followers of the world's diverse religions continues to grow. Consequently, we are more aware than ever of the variety of our earth's separate spiritual communities.

The second reason is tied closely to the first. As we indicated in the first chapter, the word *spiritual* includes two dimensions—transcendence and immanence. When it is assumed that a spiritual power stands outside the physical world, it is called transcendent. When it exists within, it is immanent. If it exists both outside of and within, it is both transcendent and immanent and can potentially intervene in history. In chapters 7–10, we described how the world religions differ in their views of these two dimensions. Also, the list of religions that appear in Figure 36 above reflect dissimilar combinations of transcendence and immanence.

This sets the stage for our third insight related to the second big question. As our expanding global village continues to foster greater interaction between the followers of the world's different faiths, we wonder if there will ever be a time in the future when only one religion will exist or when everyone will agree on a single transcendence and immanence combination? The answer is: probably not—and for the following reason.

In the first section of this chapter, we focused on the first big question of how the universe began and functions. The goal of science is to discover the laws of nature that govern the material universe. While some differences exist, after four centuries of discoveries, the vast majority of scientists accept modern science's view of the age, size, and evolution of the cosmos and life on earth. This broad agreement exists because science has developed widely accepted methods and procedures—based on empirical observations, experimentations, repetition, prediction, peer review, and so on—in its quest to discover the natural laws and their consequences. Science's self-correcting nature enables scientists from diverse fields of research to verify, modify, or falsify theories and hypotheses about how nature operates, which in turn helps build consensus.

These kinds of methods and procedures do not exist in the realm of religion. As we indicated in chapter 1, science and religion are two distinct types of human experiences, that is, two different ways of knowing. Science begins with hypotheses about cause-and-effect relationship in nature, and these are testable through inductive research procedures. Religions begin with assumptions that are not testable through scientific methods. It is not possible to demonstrate empirically that there exists or does not exist a spiritual power that is greater than the material universe that we experience with our five senses. Furthermore, each religion contains different beliefs about an Ultimate Reality that transcends materiality, is immanent within it, or is both transcendent with or without intervention and immanent.

A few examples illustrate the limitations of science in relationship to religious beliefs. Science cannot verify the existence or nonexistence of God or the soul whether defined as Brahman, Allah, Waheguru, Tao, or other. Nor can science prove or disprove such diverse religious beliefs as Christ died for the sins of the world, the Qur'an reveals Allah's perfect will, yin-yang balance leads to social harmony, and so on. On questions about the nature of life after death in the form of immortality, resurrection, reincarnation, or however defined, science is silent. Why? Because religious beliefs belong to a different kind of human experience, where assumptions about a spiritual power greater than nature are not subject to scientific analysis.

However, this does not mean that there is no connection between science and religion or that science does not influence religion and vice versa. What modern science gives to religion is an image of a vast, complex, and intricate universe whose size and age premodern scientists could not even imagine (first big question). The implication of this for religion is that the spiritual force that brought this colossal universe into being and sustains it (the second big question) is even greater and more powerful. When we combine our awareness of the age and size of the universe with an image of a spiritual force that is even greater, we are led to the next question. What is the relationship between this power and the world religions that define it so differently, as Figure 36 shows? Another way to ask this question in theological terms is: What is the relationship of the One (ultimate spiritual power) to the many (religious interpretations of it)?

As we indicated in chapter 10, the three ways by which the followers of the world religions can relate to each other are exclusivism, pluralism, and inclusivism. How do these three alternatives connect to the challenge of relating the One spiritual force that created the cosmos to the many different images by which the followers of the world religions interpret it? In answering this question, we are dealing with the fourth big question of how a higher spiritual power relates to humanity. However, before we focus on this issue, we turn next to the third big question of the relationship of the ultimate spiritual force to the material universe.

RELATIONSHIP OF SPIRITUAL
POWER TO THE UNIVERSE

There are two main issues that are connected to the third big question about how an ultimate spiritual force is related to the universe. The first pertains to what we already know as a result of modern scientific findings, and the second to what we have yet to learn.

The history of science over the past four centuries is the story of one amazing, indeed, revolutionary discovery after another, and in combination they reveal to us just how powerful the spiritual force that created and sustains the cosmos truly is. We know now that the macro space and time dimensions and the micro atomic and subatomic scales are way beyond our comprehension when compared to the kinds of activities that we engage in during our day-to-day routines. We have discovered that nature's laws are surprisingly mathematical and quantitative and that life itself is diverse and resilient.

Nonetheless, despite what we have learned, we caution against assuming that we have a full grasp of everything there is to know about the material world. When we combine the first three of the four big questions, our next step is to look to the future to increase our (1) understanding of the origin and operation of the universe, (2) the spiritual power that brought it into being, and (3) how this power relates to it. We believe that many surprises lie ahead; and if the future is anything like the past, much of what we learn will be counterintuitive to our current knowledge. For example, here are ten areas that will receive increasing attention as the future of science unfolds.

1. How did life start on earth some 3.8 billion years ago?

Modern science has no conclusive answer to this question.[4] Life started on earth about 3.8 billion years ago. No one was there to characterize environmental conditions such as temperature, rainfall, acidity of the atmosphere, energy sources (e.g., radiation from the sun and heat from under water thermal vents), chemistry of the waters, and wind and storminess of the atmosphere. There have been a variety of experiments directed at this question but, still we have no definitive results. It is close to impossible to develop an explanation (hypothesis) when so little is known about the initial conditions, the molecular possibilities, and the energy sources. This question also involves several scientific disciplines such as planetary science, biochemistry, geology, and astronomy.

2. What is dark matter?

When modern astronomers observed various galaxies, they realized that something was wrong. The spin of the stars at the edge of a galaxy was moving faster than what could be understood with the prevailing theory, force

4. McFaul and Brunsting, *God Is Here to Stay*, 67–88.

of gravity and Newton's laws of motion. The stars at the edges should have separated from their home galaxies because of the lack of gravity at these distances. They recognized that something appeared to be missing and offered two ways to explain why: (a) Some scientists considered adding to or altering Einstein's general relativity (warpage of space-time due to nearby mass or energy). This was dubious because there was an abundance of experimental evidence to support general relativity. (b) Others took a different position. In order to explain their observations, they assumed that additional but undetected matter could account for the needed extra gravity. To date, most scientists have adopted option b and call this "dark matter."

Currently, scientists have yet to produce ways to study this mysterious matter. If they could show that dark matter interacts in some way with normal matter, they would be taking a huge step forward. At this point in time, the answer to questions about the nature of dark matter is "We don't know."

3. Why is the expansion of the universe accelerating?

In 2011, the $1.5 million Nobel Prize in Physics was awarded to three astronomers.[5] They used observations and simulations (based on accepted natural laws) to show that the universe is not only expanding but the rate of expansion is accelerating. Like many earlier self-correcting discoveries, this observation was completely counterintuitive. These scientists used sophisticated, ground- and space-based telescopes to make their measurements. They also, employed powerful computers to process the images they received from advanced high-tech cameras.

In order to understand this surprising acceleration, they hypothesized the existence of a new energy in the universe—the mysterious dark energy. Today astronomers have concluded that 69 percent of the mass-energy of the whole universe consists of dark energy; and like dark matter, it is something we do not yet understand.

4. Why is there so much matter in the universe?

In 1936, American physicist Carl D. Anderson received the Nobel Prize in Physics for experimentally detecting certain particle tracks in his cloud chamber.[6] His measurements indicated that these particles had the same

---

5. Saul Perlmutter of the University of California, Berkeley; Brian P. Schmidt of Australian National University; and Adam G. Riess of Johns Hopkins University and Space Telescope Science Institute in Baltimore.

6. A cloud chamber is a particle detector used for visualizing the passage of certain

mass as an electron, but electrostatically it was positively charged with the same charge as an electron's negative charge. After a year of confirmation experiments, Anderson concluded that these anti-electrons actually do exist.[7] We now call these newly discovered particles positrons.

Just after the big bang, 13.8 billion years ago, the universe was densely populated with subatomic particles. Not all of them were normal particles or ordinary matter particles. Corresponding to each type of ordinary mass particle is an anti-particle with the same mass but with an opposite charge. For example, protons (found in the atomic nucleus) have a positive charge while anti-protons have the same but negative charge. The current scientific understanding is that the big bang produced slightly more ordinary matter particles than anti-particles by only one part in a billion. This implies that for every billion anti-particles there must have been a billion plus one ordinary particles. This implies that everything in the universe (galaxies, stars, planets, humans, and baseballs) exists because of that slight surplus of ordinary matter.[8]

What is not currently understood is why there appears to be so much more ordinary matter compared to the observed amount of anti-matter. There are several possible answers. One unproved explanation is that black holes, created just after the big bang, may have converted much of their mass and energy into ordinary matter.

5. What is the fate of the universe?

Until the 1960s, the answer for the ultimate fate of the universe was simple: the universe was in a "steady state," which means it always existed the way we observe it now, and it will always continue to exist as we currently see it.[9] In 1978, Arno Penzias and Robert Wilson received the Nobel Prize in Physics, based on their experimental discovery of what is known as the cosmic microwave background, cmb.[10] This finding was an important support for our understanding of the big bang.

subatomic particles.

7 Here is an example of the importance of multiple experiments that yield the same result. This is called "repeatability." This also shows Anderson's thoroughness.

8. This is an example of predictability in science where predictions from an accepted theory (in this case mass and energy production from the big bang) are applied to predict observed results.

9. This is an example of when something that seems to be obvious and well accepted, even back to premodern times, turns out to be not at all the case resulting from observations (interpreted using current theories, based on known natural laws).

10. This is a clear historical example of the importance of actual observations and

As stated above under number 3, dark energy comprises 69 percent of our universe and drives its expansion at an accelerated pace. We use dark energy as a placeholder name because we do not really know what it is. If dark energy continues on its current path, a *big freeze* appears to be the ultimate fate of the universe. This means that in the long run, the enlarged universe will become cold, lifeless, and dark. However, since we do not know what dark energy is, we really do not know what the ultimate fate of the universe will be. Future observational evidence will lead to confirming, modifying, or falsifying this theory about the ultimate fate of our universe and all the objects and life forms within it.

6. Does every big galaxy have a central black hole?

This story starts in the early 1960s. Astronomer Maarten Schmidt (Californian Institute of Technology) made important observations. Several of what were thought to be stars were strangely bright. Upon closer examination, he found the distances to these stars to be extremely large. The apparent brightness of these objects and their distances meant that they could not be stars. These objects were given a new name: quasi-stellar objects, or quasars. After many more astronomical observations and simulations, it was in the 1980s that astronomers began to realize what these quasars really are.

Based on observations mostly from the Hubble Space Telescope (into the 1990s), astronomers observed that black holes exist at the centers of most galaxies, including our own Milky Way galaxy. While scientists theorize about the probable origins of quasars and black holes, it is necessary to stress that the universe is so vast and complex and has been in existence for such a long time that it will take many more observations with improved technology to explain fully how these phenomena become formed.

7. Is there life in other parts of the universe?

The short answer is: We do not know.[11] If we looked for life beyond earth, how would we identify it? The most fundamental characteristics of life include these five: (a) Life needs energy to survive and reproduce. For example, heat is one source of energy for some life that might come from

---

their interpretations.

11. This is an example of how humility is necessary in modern science: We need to admit and identify what we don't know, which frequently initiates inquiries into those unknown areas.

hydrothermal vents[12] under the sea. (b) Waste energy needs to be somehow expelled from the life form. (c) Life needs water to survive and thrive. (d) Life needs to be able to adapt to its environment even if that environment changes. (e) Life needs to reproduce itself.

Our reference is but one planet. We have only one history of life's origins and diversification. To understand how life came to be so pervasive and variable on our planet, we have human intelligence and the methods of modern science. But there may be other forms of investigation that are unknown to us now.

The possibilities are so vast. Our universe has an estimated width of 92 billion light years. It has a history of about 13.8 billion years. There are at least 125 billion galaxies in the universe. Our own galaxy, the Milky Way, has about 250 billion stars. At first glance the possibilities for life are, well, nearly infinite. As of the time of this writing there are 4,197 exoplanets.[13] To say that our current ability to explore all these galaxies and star systems is limited would be a monumental understatement. Also, life forms as complex as giant redwood trees, whales, roses, ameba, eagles, dogs, and cats are likely to be rare compared to extraterrestrial life forms we might eventually detect.

8. What will happen to life on Earth?

The eventual outcome for life on Earth may depend on what happens to the Sun. This assumes that life (e.g., human life) does not self-destruct before the sun changes. Premodern civilizations worshipped the sun, recognizing its regularity and life-giving powers. The Sun itself is an ordinary star. It provides light and warmth for life here on Earth, and day and night are flawlessly regular.

However, about four billion years from now the Sun is expected to enlarge and incinerate the planets of Mercury, Venus, and possibly Earth. This outcome is expected because we have made observations of stars similar to our Sun. Also, our current theories (confirmed by observations) give us awareness of what happens, for example, at the subatomic level, how the Sun produces its energy from thermal nuclear fusion, the role gravity plays, and the pathways for energy to migrate from the Sun's core to its surface. All these observations and theories are consistent in the prediction of this outcome for the Sun and Earth.

12. Hydrothermal vents are places under the ocean surface, where water heated from under the earth's crust mixes with seawater. All kinds of ocean life have been observed that draw energy from these heat sources.

13. An exoplanet is any planet beyond our solar system.

9. Does inflation theory govern the universe?

The discovery and characterization of the cosmic microwave background, CMB, did not explain all observations and major questions. In 1981, Alan Guth of the Massachusetts Institute of Technology and three others proposed a solution to issue of multiple universes. There was a short period of expansion, called "inflation," in the early universe less than about 1/(billion, billion, billion seconds), incomprehensibly short, when all of space-time expanded by an incomprehensibly large factor. All trustworthy observational evidences confirm that inflation did happen.[14]

So, the answer to this question is probably "yes," but there are significant follow-up questions. For example: What caused inflation in the first place? What was the energy source for inflation? And, why did it stop when it did? Mysteries remain, and it is the task of science to continue the search to unveil them.

10. Do we live in multiple universes?

Since the 1990s, some modern scientists have begun to consider the possibility that there might be other universes besides our own.[15] The motivation for this idea comes from at least three sources: (a) Cosmic inflation. See number 9 above. (b) String theory. This is a theory where particles like the electron and proton are replaced by very small vibrating strings. And (c) the accelerating size of the universe. At this point in time, there is no direct evidence for multiple universes. However, there are several convincing theoretical arguments for a multiverse that will guide future research. Given the self-correcting and counterintuitive nature of many of modern science's most significant discoveries, the answer is "possibly." It in the very nature of science to continue conducting observations and experiments that might one day lead to confirming, modifying, or falsifying this theory.

As we conclude this section, we recognize that as scientists go forward with current and future observations and experiments, their discoveries will no doubt provide answers to many of the challenges identified in the above ten areas. Given the self-correcting nature of science during the past four hundred years, we can expect the unexpected or the counterintuitive findings that will once again revolutionize our views of how the universe began

---

14. Here is another example of a theory that is completely counterintuitive but still widely accepted due to its consistency with observations. For more information on inflation, see Krauss, *Universe from Nothing*, ch. 6.

15. Krauss, *Universe from Nothing*, 119, 125–29.

and operates (first big question). When we couple this with the belief that there exists a spiritual force greater than the physical world that we experience through our five senses, we will be even more amazed than we are now of the grandeur of this power (second big question). Finally, we will deepen our understanding of the yet to be discovered natural laws by which this spiritual power governs the material universe (third big question).

We are now ready for the fourth and final big question.

RELATIONSHIP OF A SPIRITUAL POWER TO HUMANITY

The fourth big question follows from the other three: How should we think about the relationship between the spiritual power that created and sustains our vast universe and humanity? In order to answer this question, we will combine various aspects of all of the previous ten chapters.

Our point of departure lies in recognizing the impact that science has had on our understanding of the spiritual power that brought our cosmos into existence and guided its evolution from the start (chs. 2–6). If this spiritual power transcends this massive material universe and/or is immanent within it, how are we perceive this power in relationship to the world's diversity of religious beliefs (chs. 7–10)? In order to answer this question, we will focus on the relationship of the One to the many, which is the central concern of this section of the chapter. In addition, we will give special attention to how Christians can understand their specific faith in relationship to the larger issue of the evolution of faith that connects Christianity to the other spiritual traditions that exists around the globe.

We start by observing that virtually all of the theistic traditions that we have described in chapters 7–10 and summarized in Figure 36 above emerged during the premodern science period, except for Baha'i, which holds that there is no conflict between science and religion. Modern non-spiritual atheism and agnosticism are reactions against traditional Western forms of monotheism as expressed in Zoroastrianism and the Abrahamic faiths of Judaism, Christianity, and Islam. Despite their atheism, the Eastern religions of Buddhism and Jainism set spiritual fulfilment as humanity's goal.

When addressing the challenge of relating the One to the many, we recognize that we are comparing a single perception of the spiritual power that created the universe against multiple views of this power, which are associated with the monotheistic, pantheistic, poly/henotheistic, and animistic faiths. As modern scientific discoveries have revealed, our universe is very old, very large, and very complex and that there are still many scientific mysteries that have yet to be solved (first and third big questions). Thus, the task of connecting the One to the many is really a challenge of how to relate the modern image of a single spiritual power that created our vast universe (the second big question) to the many different premodern perceptions of that power.

How should we think about this? In chapter 7, we described many of the changes that have occurred throughout the world in the not-too-distant past. The Guttenberg invention of the printing press in the fifteenth century stimulated an increase in literacy levels and the translation of the Bible and other sacred scriptures into multiple languages. In 1893, the leaders of different religions organized the Parliament of the World Religions and met for the first time in human history at the Columbian Exhibition—World's Fair in Chicago. Now, as a result of worldwide electronic communication and mass transportation, the earth is evolving toward greater regional out-migration and increased interaction across international boundaries. As a result, religious heterogeneity is increasing steadily through more direct contact and through access to the worldwide web where anyone can learn about other cultures and religions at the touch of a computer key. As a result of these and other trends, the earth is being transformed like never before into a growing global village. Just as the cosmos evolved from the moment of the big bang, increased interaction between the followers of the world religions is continuing to evolve.

In addition, one of the main legacies of centuries of new discoveries is that modern science has developed a systematic and self-correcting process that has improved our understanding of how our universe started and expanded. Since the big bang 13.8 billion years ago, our mammoth universe has been evolving steadily. This makes evolution one of the core images connected to the first and third big questions. From the start, evolving is what the cosmos has always done—in the past, in the present, and no doubt in the future. Simply stated: *Evolution is the process through which the spiritual power that is greater than nature created and guided the development of the universe.*

Next, when we move from the large-scale or macro level of the entire cosmos to the smaller-scale or micro level of the earth, we observe that humanity's search for Ultimate Reality has led to multiple spiritual traditions—as summarized in Figure 36. If we assume that there exists a single and ultimate spiritual power that created and is greater than our enormous universe and guided the evolution of life on earth, why do the world religions hold such different views about Ultimate Reality? The answer is clear. Their views of Ultimate Reality emerged under conditions of geographical and social isolation. Where they share common themes, as in the case of monotheism, pantheism, or other, it is because they appeared and developed variations in succession in the same geographical region, such as South Asia (Hinduism, Buddhism, and Sikhism), Central and Eastern Asia (Taoism and Shinto), or the Middle East (Zoroastrianism, Judaism, Christianity, Islam, and Baha'i).

Given that our planet is evolving in the direction of becoming an expanding global village because of electronic technology and mass transportation, previous conditions of geographical and social isolation that gave rise to religious pluralism no longer exist. More than ever in human history, the followers of the world's diverse religions are interacting with greater frequency; and unless something unforeseen occurs, this trend will only increase in the future.

Assuming that greater interfaith contact continues, is it likely that the followers of diverse religions will evolve a shared view of the spiritual power that created and guided the evolution of the universe in the same way that most modern scientists have developed a common perception of the origin and evolution of the universe? The answer is probably not. Thus: *As the global village continues to expand in the future, the evolution of faith will parallel this process through a growing awareness of and response to religious pluralism.*

To conclude, as the above discussion of the four big questions shows, the two main areas that cut across all ten of the preceding chapters are evolution and religious pluralism. How are these related to Christianity? We will answer this question in the next and final chapter.

12

The Evolution of Faith: Christianity and Evolutionary Pluralism

INTRODUCTION

Evolutionary Pluralism (EP) is the name that we are giving to the belief that evolution applies to both science and religious pluralism and that Christianity's relationship to both can stimulate the evolution of faith and greater interfaith harmony. Since our focus is on Christianity, we will use the name by which Christians identify Ultimate Reality: God. We recognize that in the search for Ultimate Reality, the world religions that we describe in this book have developed multiple names and viewpoints. This does not mean that the various vocabularies that different religions use to define the Ultimate are inferior to Christian terminology, only that names for Ultimate Reality differ depending on how languages and cultural traditions vary from society to society. Thus, when we describe how Christianity relates to Evolutionary Pluralism and the evolution of faith, we will use the word God to mean Ultimate Reality or the spiritual power that created the universe and beyond which nothing is greater.

CHRISTIANITY AND MODERN SCIENCE

Our understanding of the universe radically changed from premodern times to modern times. In other words, just as the universe evolved over time, so has our understanding of how this occurred. In those ancient times the flat earth was the center of it all with the sun, moon, stars all moving around our planet during a twenty-four-hour cycle. Without telescopes, this is exactly the way the heavens appear. The image for this was an inverted bowl over a flat earth. (See ch. 2, fig. 4.) A *Prime Mover* was thought to have caused this motion. Those movements above the earth were thought to be in perfect, untainted circles. The space above the earth was thought to be perfect, uncontaminated, and pure. By contrast the space on the earth was imperfect, contaminated, and impure. Humankind was at the center of the universe on earth's surface.

It is hard to see how this premodern view could be any more different than what we understand today. Our current perception is that our earth is just one planet of eight that orbit our sun. Our planet earth has only an average size. Also, we now know that our sun is a rather ordinary star in the outer regions of our galaxy, the Milky Way. Vast spaces separate us from the other planets, stars, and galaxies. The Milky Way itself has about 200 billion stars (plus or minus 100 billion stars). Our Sun is but one of those 200 billion stars. Our galaxy is but one of at least 150 billion galaxies in the universe that we can detect.

A mere 120 years ago our universe was understood to be very dissimilar compared to how we see it today. At that time the Milky Way was thought to be the whole universe. There was nothing else beyond our galaxy. All the other 150 billion or so galaxies were unknown at that time. The stellar disc of the Milky Way is about 200,000 light years in diameter. Today the diameter of the known universe is estimated to be 92 billion light years. (See ch. 11, fig. 35.) One hundred twenty years ago the universe was thought to be static and unchanging. It was thought that the night time sky that we observe now would have been similar to what one would see far back into the infinite past and far into the infinite future.

Today our understanding is much different regarding change. Our current theory is that the big bang occurred about 13.8 billion years ago when the entire universe was located within an infinitely dense, infinitely small point, and infinitely hot tiny space called a singularity. From that totally incomprehensible condition, the universe evolved into subatomic particles (for example, protons, neutrons, quarks, and photons), into stars, into galaxies, into planets, and about 200,000 years ago into us. As if this were not enough, all of space and time evolved from 13.8 billion years ago from that infinitely small singularity into the space and time we live in today. The question "What happened before the big bang?" cannot be answered because apparently "before" the big bang, both time and space did not exist.

When we shift from the above modern scientific description of the enormous age and size of our universe (first big question) to an assumed existence of a spiritual reality that is even greater (second big question), we find ourselves asking the following question: How can Christians, as well as the followers of other faiths, understand the main characteristics of this extraordinary power? Our answer consists of three parts.

First, given that the size of the universe is 92 billion years wide, the spiritual power that is greater must have a spatial presence that exceeds this width; second, given that space-time started 13.8 billion years ago, this power must have a temporal presence that exceeds this time; and third, given that modern science has a unifying theory in particle physics called the Standard Model that leads to making accurate predictions about natural phenomena, the spiritual power that is greater than materiality must be involved in the most fundamental aspects of physical evolution where seventeen components of nature (Higgs boson, six leptons, four force-carriers, and six quarks) form its basis.

In light of these three characteristics, we conclude that God who created the universe and guided its expansion is incomprehensibly vast in space, time, and involvement with the physical evolution of the universe.

CHRISTIANITY AND GOD

What is the main lesson of the above summary of modern science's view of the origin, age, and size of the universe for the evolution of faith and Evolutionary Pluralism? From the perspective of Christianity, the supreme spiritual power or God who created the universe surpasses any and all of the multiple images that the world religions have created about the nature of Ultimate Reality—including the description of God that appears in the biblical story of creation. This does not mean that the various interpretations of God or Ultimate Reality as envisioned in other religions are invalid because of their dissimilarities. It means only that they are limited in grasping God's essence in its fullness, including Christianity.

We have made this abundantly clear in chapters 7–10 where we have shown how the world's religions differ in their diverse combinations of the four factors of materiality, spirituality, transcendence with or without intervention, and immanence in the forms of monotheism, pantheism, polytheism/henotheism, animism, and atheism/agnosticism. Furthermore, we hold that we cannot use the empirical approaches of modern science (experimentation, observation, prediction, peer review, and so on) to determine if one of the many images of God is correct in the same way that science can verify, modify, quantify, or falsify a scientific hypothesis through its self-correcting methods. Science and religion are different ways of knowing. The methods that apply to discovering the causal relationships governing nature that we experience with our senses do not apply to the religious realm that involves assumptions about a spiritual power that created nature and is transcendent with or without intervention and/or immanent within it.

It is precisely at this point that we become acutely aware of one of the major paradoxes on modern science in its relationship to religion. While science cannot verify the existence of a spiritual power that is greater than nature, the discoveries of modern science imply that we live in an unbelievably old, large, and continuously evolving universe. While many might think that this modern view of the universe undermines belief in God, the paradoxical outcome is the opposite: Modern science appears to support this belief by revealing its finely tuned structures. The next step is easy to take from the perspective of Evolutionary Pluralism and the evolution of faith. When we shift from the first to the second big question, the world's religious images of Ultimate Reality, by whatever name, are incomplete glimpses of the full essence and purpose of this extraordinary spiritual power that created the cosmos.

In addition to this paradox, modern science teaches us not only that the presumed God or Ultimate Reality is greater than both the universe and the limited views of the world religions, but the impact that science has on religion is counterintuitive. How so? In chapters 2–6 we described how many of the most significant discoveries and theories of modern science went contrary to the accepted scientific beliefs or intuitions that once existed at the time, including those of Galileo, Kepler, Heisenberg, Einstein, Hubble, Darwin, and others. As a result, our understanding of how the universe started and developed has not remained static during both the premodern science and modern science eras.

When we turn to the realm of religion, all of the images of God or Ultimate Reality were created during the premodern science era with the exception of Baha'i that began in the mid-nineteenth century. (See ch. 8, Table 10.) For Evolutionary Pluralism, the counterintuitive impact that modern science has had on religion is that it presents a vision of God or Ultimate Reality that is greater than any of the world religions' premodern socially created and therefore limited perspectives. In the language of theology, here is the key point: the counterintuitive impact that modern science has had on religion—including Christianity—is that it has broadened our understanding of God. This means that the real challenge that Christians, as well as the followers of the other world religions, encounter is to relate this broader modern scientific vision of a single powerful and creative force to the many narrower premodern interpretations of it. Just as modern science continues to evolve, so does our understanding of the God who created and guided the evolution of the universe. Evolution applies both to science and faith. In other words, building on the paradoxical and counterintuitive impact that modern science has had on religion, the challenge Christians face, as well as the followers of other faiths, is how to relate modern science's implied understanding of the One to the premodern views of the many?

INERRANCY AND SCRIPTURE

In order to answer this question of the relationship of the One Ultimate Reality, or God, to its many interpretations, it is essential that we turn our attention to the scriptures of the world religions, which provide us with multiple perspectives and images of the Ultimate. What is clear from the outset is that all of these sacred texts were written by multiple authors who lived under very different situations involving diverse cultures and languages. These scriptures are the primary sources through which the followers of the world's diverse religions integrate materiality, spirituality, transcendence with or without intervention, and immanence into substantially different combinations.

As we have shown in chapters 7–10, the Zoroastrian, Jewish, Christian, Muslim, and Baha'i scriptures present monotheistic images of the Ultimate.[1] Hindu, Taoist, and Sikh scriptures differ in giving us pantheistic visions.[2] Ancient Greek and Roman mythology[3] as well as indigenous oral traditions point to patterns of polytheism and henotheism. Japanese texts provide us with an animistic form of spirituality.[4] Buddhist and Jain sacred books[5] combine atheism with the goal of spiritual fulfilment, whereas Western atheist and agnostic writers emphasize the nonexistence of any form of spirituality that is greater than materiality.[6]

1. Respectively, they are: Avesta, Tanakh, Holy Bible, Qur'an, and Book of Certitude.

2. Examples include the Bhagavad Gita and the Upanishads, Tao Te Ching, and Adi Granth.

3. Greek: Zeus and numerous others; Rome: Jupiter and numerous others taken from the Greeks and renamed.

4. Kojiki and Nihongi.

5. Examples include the Dhammapada, the Lotus Sutra, and the Agama.

6. Two examples of writings from the movement called New Atheism include

This takes us to the next question. What is the relationship between science's implied image of a single amazing spiritual power and Christianity's interpretation of how this power is related to the creation and evolution of the universe and of faith? In chapter 5, we discussed the importance of the doctrine of biblical inerrancy as one of the key assumptions related to the four Christian perspectives of Young Earth Creationism (YEC), Old Earth Creationism (OEC), Intelligent Design Creationism (IDC), and Evolutionary Creationism (EC). Table 6 shows how these four positions differ.

When we compare them to each other, we observe that there exists a sliding slope in how they apply biblical inerrancy to modern scientific discoveries. For example, YEC applies inerrancy to a literalist interpretation of the entire Bible. OEC and IDC modify inerrancy to accommodate the modern scientific view of the physical age and size of the universe but not biological evolution. EC slides farther down the slope by accepting not only modern science's image of the cosmos but also evolution as the process by which life on earth developed. In other words, the four views of YEC, OEC, IDC, and EC themselves show an evolutionary pattern that changes from the biblical literalism that rejects the modern scientific view of evolution to complete acceptance of it.

Thus, it appears that those who employ the doctrine of biblical inerrancy to reconcile the differences between modern science's view of the universe and the Genesis account of creation end up with multiple interpretations that start with complete acceptance of the biblical narrative (YEC), go next to partial acceptance (OEC and IDC), and end with complete rejection (EC). Evolutionary Pluralism (EP) accepts along with Evolutionary Creationism (EC) that the modern scientific perception of the origin and evolution of the cosmos has thus far been shown to be accurate. Therefore, this perception supersedes all premodern perspectives including YEC's literal and OEC's and IDC's partial interpretations of the Bible's creation narrative.

Evolutionary Pluralism (EP) does not include a doctrine of scriptural inerrancy. It understands that the sacred texts of all the world's religions are historically conditioned and limited in their understanding of God and God's relationship to the universe and humanity—including Christianity. This means that none of the world scriptures is inerrant in full or in part. At the same time, EP accepts that the world's sacred texts are divinely inspired, which means that their authors received spiritual insights.

Dawkins, *The God Delusion*, and Stenger, *God: The Failed Hypothesis*.

INERRANCY AND SALVATION

While the advocates of YEC, OEC, IDC and EC disagree over how to apply the doctrine of inerrancy to the biblical story of creation and evolution, there is one area where they are in complete agreement: the Bible is inerrant in claiming that the only way to salvation is through Christ.

Where does Evolutionary Pluralism stand on this issue? EP does not accept that there is only one pathway to salvation. We define the doctrine of salvation—or soteriology—to mean the anticipation of a positive life after death. It can also refer to earthly experiences of what this afterlife condition might be like. For Christians, this is often called realized eschatology, which refers to Christ's spirit of love that is present among us while we are alive. However, no one knows or can verify with scientific certainty what life after death entails. From the perspective of the world's diverse spiritual communities there are multiple visions, such as biblical view of the resurrection of the dead, the Greek image of the immortality of the soul, the Eastern belief in reincarnation, indigenous religious assumption that the souls of the departed join with the souls of deceased love ones in spirit world, and so on.

For Christians, the expectation of achieving a blessed eternal life after death is coupled with the doctrine of Christ's redeeming sacrifice. At the same time, Evolutionary Pluralism and the evolution of faith accepts that there exist multiple spiritual pathways to a hope-filled positive afterlife as the followers of the world's diverse religions express them. Thus, EP not only rejects biblical inerrancy when it is applied to the Genesis account of creation but to the exclusivist perspective that there is only one way to achieve a rewarding life after death. This is consistent with modern science's implied broader view of God who is greater than all of the premodern, finite, and narrower scriptural interpretations of God's nature and purpose in this life as well in a potential life to come after death.

At this point, we can combine this broad view of God with the global demographic distribution of the world religions. Currently, Christianity is the largest religion on earth at 32 percent or about one-third of the world's population. The remaining 68 percent or slightly more than two-thirds belong to other religions or remain nonaffiliated with any religion. If the only pathway to salvation goes through Christ, and if the large majority of people on the planet are not Christian, (about two out of three), then it means that the majority of the people on earth cannot be blessed with the gift of God's grace as it applies to salvation. See the following bar chart.

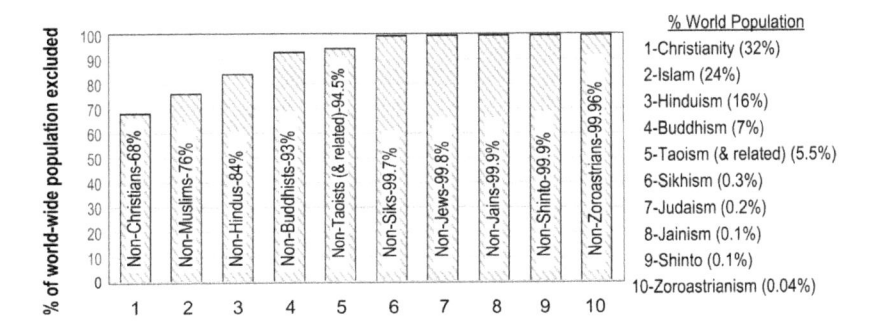

Figure 37. Bar chart of the percentage of people excluded from salvation by religion.

Furthermore, if Christianity can claim that Christ is the only means to salvation, then the followers of other religions can make a similar claim that only their faith offers a pathway to salvation or Enlightenment as in the case of the Eastern religions. For Hindus, it is through Brahman; for Buddhists, Buddha; for Jews, obeying the Torah; for Muslims, obeying the Qur'an and Muhammad; for Taoists, balancing the yin and yang, and so on. Each of these views represents a premodern, limited perception of the relationship of Ultimate Reality to humanity. Each religion's salvation pathway applies only to its followers and not to the members of any other spiritual community.

However, if we assume that there is a truly amazing God who brought our incomprehensibly old and large cosmos into being and guided its evolution, then we must go beyond the historically conditioned and limited worldviews that define pathways to salvation through premodern, restrictive doctrines. Given that the world religions developed in different societies and cultures with different languages, it is understandable why their soteriological doctrines, that is, what happens in the afterlife, would differ in the same way in which their premodern scientific views of creation vary.

Just as modern science's perception of our huge universe surpasses all premodern limited perspectives, the broad view of the God who created this enormous universe surpasses all premodern narrow conceptions of God. This points toward a God who is the God of all humanity and not just a select group and that God's salvation is potentially available to everyone. Thus, in keeping with the evolution of faith, this means that amid the many spiritual communities that arose out of very different historical contexts and that currently populate our planet, we can readily accept that God offers multiple pathways to salvation. Furthermore, like the parable of the elephant and the blind, each religion contributes a partial and incomplete

understanding of the Ultimate and the pathway to salvation or enlightenment through its own unique spiritual experiences.

From the perspective of Christianity, it would seem inconsistent to
replace the premodern biblical understanding of creation with the modern
scientific view of evolution and still claim that there is only one pathway to
a positive life after death as defined by the very scripture whose premodern
view of creation is either modified to accommodate modern science in part
(OEC and IDC) or is rejected in full (EC). Why disavow the doctrine of
biblical inerrancy in the case of creation but not salvation? Does this not
imply a selective application of the doctrine of biblical inerrancy without
specifying the criteria for full or partial acceptance of the creation story or
for rejecting it completely?

Just as we have progressed in our perception of the age and size of
the universe as a result of modern science, we also need to progress in our
view of God's eternal promise that follows from an understanding of the
powerful, creator God that modern science implies. Along with Young
Earth Creationism, Old Earth Creationism, Intelligent Design Creationism,
and Evolutionary Creationism, Evolutionary Pluralism assumes that Christ
is Christianity's pathway to salvation. However, where EP differs from the
other four is in rejecting that Christ is the only way. Given that the Christian
faith can be integrated readily with the modern scientific understanding of
the evolution of the universe, it can also be accommodated to the belief that
other religions offer legitimate and equally valid positive pathways to life
after death. This means that even as the many scriptures of the world religions provide a diversity of images of the Ultimate Reality that Christians
call God, they also define alternative means of Salvation or Enlightenment
that for Christians is grounded in Christ.

QUEST FOR THE ULTIMATE AND COMMON ETHIC

Next, we turn our attention to a seeming incongruity that cuts across all the
world religions. On one hand, we have shown that the search for the Ultimate has led to a wide range of images of Ultimate Reality that exist among
different religious communities. On the other hand, in chapter 7, we noted
that there exists in all the world religions a universal reciprocity norm also
called the golden rule. The human inclination to do good to others because
we want them to do good to us or to not harming others because we do not
want them to harm us is universal. It is found in all cultures.[7]

7. For an extended discussion of how the universal reciprocity norm or golden rule
appears in all the major religions of the world as well as in philosophy, see McFaul and

At the same time, this universal ethical norm is all too often discarded or suppressed entirely during times of religious or other forms of cultural, political, military, or personal conflicts. The history of warfare that has occurred in all societies makes this abundantly clear. Nonetheless, the golden rule abides in the collective human conscience; and when hostilities recede, it reasserts itself as the basis for a return to social harmony and cooperation despite the ongoing differences that persist between different groups. Another way to say this is that just as hostilities between warring groups destabilize society, the resurgence of the universal reciprocity norm helps to keep them stable or to restabilizes them.

In the evolution of faith, the challenge that Christians and the followers of all the religions face is how to enhance moral mutuality with others in the midst of persistent theological and philosophical differences that push in the direction of increasing distrust and misunderstanding. Religious worldviews separate societies from each other, whereas the golden rule unites them. Thus, the perennial tension: in a world of fragmented religious traditions—monotheism, pantheism, polytheism/henotheism, animism, and so on—that push us apart, the golden rule cuts across all of them to pull us together. The challenge that Christians and the followers of other spiritual communities confront is how to keep theological and philosophical dissimilarities from escalating into hostilities and disruptions that destroy the potential for moral reciprocity that serves to enhance human well-being across the boundaries that keep groups separated.

All the scriptures of the world religions contain passages that tie together a vision of the Ultimate with moral mandates that are designed to promote peaceful behavior. For Christians, this is, of course, the Bible that contains numerous such references in both the Old and New Testaments. For example, in his Sermon on the Mount as recorded in the fifth chapter of the Gospel of Matthew, Jesus expresses his well-known list of blessings known as the Beatitudes. In Matt 5:9, he says, "Blessed are the peacemakers, for they will be called children of God."

In addition, in Luke's Gospel as we indicated in chapter 10, a lawyer asks Jesus what he must do to inherit eternal life. In response, Jesus ties the love of God to the universal reciprocity norm or golden rule by quoting from Deut 6:5 and Lev 19:18. "You shall love the Lord your God with all your heart, and with all you soul, and with all your strength, and with all your mind; and your neighbor as yourself" (see Luke 10:27). The lawyer follows up with a second question, "And who is my neighbor?" Jesus responds with his parable of the Good Samaritan who pitied and cared for an injured traveler who was

---

Brunsting, *God Is Here to Stay*, 144–49.

robbed, beaten, and left half-dead while traveling from Jerusalem to Jericho. The biblical message is clear. In the name of God's love, the Good Samaritan saw the downtrodden traveler as a neighbor in need and helped him.

Jesus's parable of the Good Samaritan carries special significance during our current era of the expanding global village where interaction between the followers of different world religions is growing steadily. The parable is an Evolutionary Pluralism story that includes both Judaic and Samaritan elements. The Samaritan is an outsider from the northern region of Samaria; and he is journeying in the southern land of Judea from the main city of Jerusalem to Jericho, which are two of the most important Jewish cities. Thus, the Good Samaritan narrative is consistent with the evolution of faith that stretches beyond limited boundaries. When we combine Jesus's references to the blessing of peacemaking and the love of God and neighbor with the Good Samaritan parable, we discover one of the core messages of the New Testament. What do Jesus's teachings imply for Evolutionary Pluralism and the evolution of faith in how Christians should relate to non-Christians who are becoming increasingly the neighbors next door?

CHRISTIANITY AND OTHER FAITHS

In chapter 10, we described exclusivism, pluralism, and inclusivism as the three alternatives associated with how the members of any given religion might relate to the followers of other faiths. Exclusivists posit that only they possess the truth. Inclusivists identify their beliefs as supreme among the world religions and measure others' beliefs by their own standards. Pluralist assume that all religions are equal but limited in their understanding of the spiritual power that created but is greater than the universe. From the perspective of the evolution on faith, which of these three alternatives is the preferred Christian choice and why?

The answer is pluralism for the following two reasons. One is theoretical, and the other is practical. On the theoretical side, it is highly unlikely that the disagreements between Christians over how to understand the relationship between Christianity and non-Christian religions can be overcome so that everyone would hold the same view. Differences between Christian exclusivists, Christian pluralists, and Christian inclusivists will continue. Christians disagree over Christology as well as the sources of authority that support any given view of Jesus and his ministry and mission. Theological differences existed from the dawn of the church and continue down to this day; and no doubt they will continue well into the future unless something totally unexpected occurs to change this—which really is not very likely to happen.

Thus, it appears that the issue of how Christians relate to non-Christians as well as to Christians who hold different views comes down to the practical aspects of living the faith in daily encounters with others both inside and outside of the church. Given that there is no consensus among Christians on choosing exclusivism, pluralism, or inclusivism as the preferred theological choice, we are left to ask: Which of these positions is most likely in practice to contribute the most to increasing the prospect of peace as Jesus expressed it in his Beatitudes and response to the lawyer to love God with all your heart, soul, strength, and mind and your neighbor as yourself?

On this question, the answer seems clear. Pluralism is more likely than the other two to cultivate peaceful relationships between Christians and non-Christians. We can rank order these three alternatives in terms of their peace-evolving possibilities. Pluralism ranks highest on the list because it carries more potential than either inclusivism or exclusivism for fostering peace in the global village. Inclusivism is second, and exclusivism is third. The Christian exclusivist option is the narrowest and least capable of inspiring pluralistic collaboration because its adherents hold that God's saving grace can come only through Christ.

Since Christianity is the largest religion, an exclusivist interpretation of the Christian view of salvation would exclude nearly two-thirds of the world's people, as we have shown above in Figure 37, "Bar chart of the percentage of people excluded from salvation by religion." The devout followers of non-Christian faiths will no doubt reject this narrow view of salvation, and many will define it as arrogant and patronizing. The same holds true for non-Christian religions, but even more so. All of the non-Christian faiths have fewer members than Christianity. If the followers of any non-Christian faith assume that their view provides the only pathway to eternal life or a positive life after death however defined, then even more people would be excluded from this blessing than would be excluded under Christian exclusivism, as the bar chart shows.

Inclusivism falls between exclusivism and pluralism, which effectively makes it a compromise. Inclusivism moves beyond exclusivism by accepting that God's grace embraces both Christians and non-Christians. However, by viewing the spiritual experiences of non-Christians through the theological lens of Christ who fulfills perfectly the divine mystery toward which all religions strive, inclusivism stops short of the pluralist position that all are equal under God's grace, even though God's grace encompasses all of them as we have shown in our discussion of Rahner in chapter 10. Whether intended or not, this undermines the integrity and distinctiveness of non-Christian experiences. Thus, at the practical level in terms of which of the three positions comes closest to promoting Jesus's image of the peacemakers, whom he calls the children of God, it is pluralism that possesses the greatest potential.

EVOLUTION OF FAITH, EVOLUTIONARY PLURALISM, AND THE BIBLE

For Christians, there is ample biblical support for Evolutionary Pluralism as an essential next step in the evolution of faith. At the same time, as we have demonstrated throughout this book, the Bible is subject to multiple interpretations. In chapters 5, we identified the four positions of Young Earth Creationism (YEC), Old Earth Creationism (OEC), Intelligent Design Creationism (IDC), and Evolutionary Creationism (EC) as alternative applications of the doctrine of biblical inerrancy to creation. Then, in chapter 6, we compared all four against the guidelines that the National Academy of Sciences has established for conducting modern scientific research. Of these four positions, Evolutionary Creationism is the only one to accept in full the modern scientific view of the universe. Evolutionary Pluralism (EP) and EC are in complete agreement on this issue.

Thus, for Evolutionary Pluralism, modern science and Christian faith are completely compatible. They are mutually reinforcing and grounded in one of the Bible's key theological assumptions, namely, that God created the universe; and it is modern science that gives us the most accurate and reliable picture of how this happened. Like Evolutionary Creationism, Evolutionary Pluralism has no need to retain a premodern view of the origins of the cosmos by appealing to the doctrine of biblical inerrancy or divine command of the origin of species. Letting go of these perspectives represents a major step forward for Christianity in the twenty-first century and beyond. This means that the modern scientific view of the universe is no threat to the Christian faith. Instead, the opposite is true. The discoveries of modern science (for example, regarding the age and size of the universe and the diversity of life) reinforce the Christian understanding of just how great and amazing the creator God really is.

In addition, just as the doctrine of biblical inerrancy can be interpreted in different ways or even eliminated altogether when applied to the Genesis story of creation, it can also be discarded when applied to an exclusivist doctrine of salvation. Evolutionary Pluralism does not accept that Christ is the only pathway to anticipating a positive after life. Instead, Evolutionary Pluralism, differs from Young Earth Creationism, Old Earth Creationism, Intelligent Design Creationism, and Evolutionary Creationism by eliminating biblical inerrancy in the case of both creation and salvation—for this reason. Eliminating the doctrine of biblical inerrancy is not the same as eliminating the significance and centrality of the Bible and Christ for Christianity and the evolution of faith.

Not only is the Bible compatible with modern science's implied view of the powerful God who created our vast universe, it is also compatible with the view that God's grace is equally available to all in both life and death, however envisioned. What is incompatible with the Bible is the presumption that God's grace does not apply to nearly two-thirds of the world's population that is not Christian. Such an exclusivist view implies that only Christians can claim to know God's mind in defining who is worthy of God's eternal grace that includes the possibility of some form of blessed life after death as well as a partial experience of what this is like while alive.

Such a restrictive view goes contrary to the Bible's broad understanding of God as revealed in Christ. Apart from some selective exclusivist passages and narrow interpretations of them, the Bible's core message is this: God's grace and love, as made known to us through the life and mission of Jesus of Nazareth, incorporates all humanity—that is to say, every person irrespective of their culture, place in society, or religious heritage. This understanding lies at the core of Evolutionary Pluralism and is consistent with the evolution of faith in the twenty-first century and beyond.

This means that God's grace embraces not only Christian monotheists, but non-Christian monotheists as well, along with pantheists, polytheists/henotheists, animists, and—yes—even atheists and agnostics. In the universal search for Ultimate Reality, out of their unique historical circumstances and cultural experiences, all of the world's diverse religions developed their own theological and philosophical conceptions of the Ultimate. Now, as a result of modern electronic communication and mass transportation, we are aware of humanity's spiritual diversity more than even before in human history. The real challenge that Christians confront, along with the followers of all religions, is to evolve an understanding of faith that leads to greater cooperation and peaceful coexistence across the boundaries of all faiths.

From the perspective of Evolutionary Pluralism, modern science helps point the way toward this goal by starting with the assumption that a powerful spiritual force greater than materiality brought the universe into being 13.8 billion years ago through the big bang and guided its evolution. Modern science's implied vision of God is consistent with the biblical assumption that God is Lord of the entire universe; and this encompasses all of the physical, social, cultural, and spiritual forms of diversity that have evolved on earth. Like the discoveries of modern science, God's abundant blessings are spread across the landscape of religious pluralism in this life as well as in anticipating a positive life after death.

CONCLUSIONS

The perspective that we have developed in this chapter and throughout this book starts with the four big questions that we listed in chapter 1. In chapters 2–11, we have answered these questions by focusing on the amazing and mostly counterintuitive discoveries of modern science and the even more amazing God who created the universe and guided its evolution. Through interfaith comparisons of how different religions combine materiality, spirituality, transcendence with or without intervention, and immanence, we have shown that no one religion embodies the totality of spirituality through its history and traditions, including its sacred texts. The God of the cosmos is also Lord of humanity in all its diversity as represented in the multiple spiritual communities that comprise the world of religious pluralism.

Furthermore, in the midst of the variety of faiths that populate our planet as a result of the search for the Ultimate, a universal reciprocity norm called the golden rule cuts across all of them. Unlike the theological and philosophical dissimilarities that divide the followers of world religions, the golden rule helps unite them. It is a force for unity amid diversity. Assuming that there is a God who designed, created, and sustains this universe, then surely it must be true that the same God has inspired multiple world religions and implanted the universal reciprocity norm and the search for the Ultimate in the believers of those religions.

The view of the evolution of faith that we develop throughout this book includes the universal reciprocity norm that is reflected in the words of Christ, who teaches us to love God fully in every regard and others as ourselves and to follow in the footsteps of the Good Samaritan by caring for the needy and disadvantaged. The God we see expressed through Christ is consistent with the mighty and amazing Creator God implied in modern science as well as the loving and compassionate God whose grace encompasses the followers of the world's pluralistic spiritual communities. In biblical terms, "God is love, and those who abide in love abide in God, and God abides in them" (1 John 4:16) and "Beloved, let us love one another, because love is from God; everyone who loves is born of God and knows God" (1 John 4:7).

Thus—to conclude this book—from the perspective of the evolution of faith, it is the combination of modern science and religious diversity based on the biblical understanding of God's universal love as revealed through the life and mission of Jesus Christ that we are calling Evolutionary Pluralism.

Bibliography

Baha'u'llah. *The Book of Certitude*. Translated by Shoghi Effendi. Wilmette, IL: Baha'i Trust, 2003.

Barbour, Ian. *Religion and Science: Historical and Contemporary Issues*. San Francisco: Harper, 1997.

Bauer, Susan Wise. *The Story of Western Science: From the Writings of Aristotle to the Big Bang Theory*. New York: Norton, 2015.

The Bhagavad Gita. Translated by Juan Mascaro. New York: Penguin, 1962.

Bortz, Fred. *Charles Darwin and the Theory of Evolution by Natural Section*. New York: Rosen, 2014.

Bortz, Fred. *Laws of Motion and Isaac Newton*. New York: Rosen, 2014.

Bryson, Bill. *A Short History on Nearly Everything*. New York: Broadway, 2003.

Challoner, Jack. *Exploring the Mysteries of Genius and Invention*. New York: Rosen, 2017.

Clayton, Philip. *The Oxford Handbook of Religion and Science*. New York: Oxford University Press, 2008.

Cohen, I. Bernard. *Franklin and Newton*. Philadelphia: American Philosophical Society, 1956.

Collins, Francis S. *The Language of God*. New York: Simon & Schuster, 2006.

Currie, Steven. *The Importance of Evolution Theory*. San Diego: Reference Point, 2016.

Darwin, Charles. *The Descent of Man*. London: Penguin Classics, 2004.

———. *On the Origin of Species by Means of Natural Selection, or the Preservation of Favored Races in the Struggle for Life*, 1st ed., London, England: John Murray, 1859.

Darwin, Francis, ed. *Darwin: His Life Told in an Autobiographical Chapter, and in Selected Series of His Published Letters*. London: Murray, 1908.

Dawkins, Richard. *The Greatest Show on Earth: The Evidence for Evolution*. New York: Simon & Schuster, 2009.

———. *The God Delusion*. New York: Bantam, 2006.

The Dhammapada: The Essential Teaching of the Buddha. Translated by Friedrich Max Muller. London: Watkins, 2006.

Durkheim, Emile. *The Elementary Forms of the Religious Life*. Translated, with introduction, by Karen E. Fields. New York: Free Press, 1995.

Eck, Diana L. *A New Religious America: How a "Christian Country" Has Become the World's Most Religiously Diverse Nation*. New York: HarperCollins, 2001.

Fesko, V. F. *The Covenant of Redemption: Origins, Development, and Reception*. Gottingen, Germany: Vandenhoeck & Ruprecht, 2015.

Feynman, Richard P., et al. *The Feynman Lectures on Physics*. 3 vols. Reading, MA: Addison-Wesley, 1963.

Freud, Sigmund. *The Future of an Illusion*. New York: Norton, 1989

———. *Moses and Monotheism*. New York: Knopf, 1939.

Hawking, Steven. *Brief Answers to the Big Questions*. New York: Random House, 2018.

———. *On the Shoulders of Giants: The Great Works of Physics and Astronomy*. Edited with commentary by Stephen Hawking. Philadelphia: Running, 2004.

Hick, John. *An Interpretation of Religion: Human Responses to the Transcendent*. New Haven: Yale University Press, 1989.

Hubble, Edwin. *The Realm of Nebulae*. New Haven: Yale University Press, 1937.

Johnson, Todd M., and Brian J. Grim, eds. *World Religion Database*. Leiden: Brill, 2008.

———. *The World's Religions in Figures: An Introduction to International Religious Demographics*. Oxford: Wiley-Blackwell, 2013.

Kant, Immanuel. *Groundwork of the Metaphysics of Morals: A German-English Edition*. Edited by Mary Gregor and Jens Timmermann. Cambridge: Cambridge University Press, 2011.

Krauss, Lawrence M. *A Universe from Nothing: Why There Is Something Rather Than Nothing*. New York: Atria Paperback, 2012.

Kroll, Kathleen. *Isaac Newton*. New York: Viking, 2006.

Krukonis, Greg, and Tracy Barr. *Evolution for Dummies*. Hoboken, NJ: Wiley, 2008.

Launiu, Roger D., and Andres K. Johnson. *Smithsonian Atlas of Space Exploration*. New York: HarperCollins, 2009.

The Laws of Manu. Translated by Wendy Doniger. New York: Penguin, 1991.

Leeming, David A. *Creation Myths of the World: An Encyclopedia*. 2nd ed. 2 vols. Santa Barbara, CA: 2010.

Leeming, David A., and Margaret Leeming. *A Dictionary of Creation Myths*. New York: Oxford University Press, 1995.

Love, David K. *Kepler and the Universe: How One Man Revolutionized Astronomy*. New York: Prometheus, 2015.

Marx, Karl. *Marx on Religion*. Edited by John Raines. Philadelphia: Temple University Press, 2002.

McFaul, Thomas R. *The Future of Peace and Justice in the Global Village: The Role of the World Religions in the Twenty-First Century*. Westport, CT: Praeger, 2006.

McFaul, Thomas R., and Al Brunsting. *God and Randomness*. Eugene, OR: Wipf & Stock, 2017.

———. *God Is Here to Stay: Science, Evolution, and Belief in God*. Eugene, OR: Wipf & Stock, 2014.

Nash, Darren. *Evolution in Minutes*. London: Quercus, 2017.

National Academy of Sciences. *Science and Creationism: A View from the National Academy of Sciences*. 2nd ed. Washington, DC: National Academy of Sciences, 1999.

Newman, James R. *Science and Sensibility*. New York: Simon & Schuster, 1961.

Newton, Isaac. *The Principia: The Authoritative Translation and Guide; Mathematical Principles of Natural Philosophy*. Translated by I. Bernard Cohen et al. 1st ed. Berkley: University of California Press, 1999.

Nye, Bill. *Undeniable: Evolution and the Science of Creation*. New York: St. Martin's, 2014.

Pew Research Center. "The Future of World Religions: Population Growth Projections, 2010–2050." Pew, April 2, 2015. https://www.pewforum.org/2015/04/02/religious-projections-2010-2050/.

English Translation of the Message of the Qur'an. Translated by Syed V. Ahamed. 2nd ed. Lombard, IL: Book of Signs Foundation, 2006.

Rahner, Karl. *Christianity and Other Religions*. Edited by John Hick and Trian Hebblethwaite. Glasgow: Fount, 1980.

Ranney, Wayne. *Carving Grand Canyon: Evidence, Theories, and Mystery*. Grand Canyon, AZ: Grand Canyon Association, 2012.

Rosling, Hans, et al. *Factfulness: Ten Reasons We're Wrong about the World—and Why Things Are Better than You Think*. New York: Flatiron, 2018.

Schoub, Barry David. *Seeing God through Science*. Eugene, OR: Wipf & Stock, 2019.

Scientific American. *Evolution: A Scientific American Reader*. Chicago: University of Chicago Press, 2006.

Shermer, Michael. *Skeptic: Viewing the World with a Rational Eye*. New York: Holt, 2016.

Spilsbury, Louise. *Johannes Gutenberg and the Printing Press: Inventions That Changed the World*. New York: Rosen, 2016.

Steele, Philip. *Galileo: The Genius Who Faced the Inquisition*. Washington, DC: National Geographic, 2005.

Stenger, Victor J. *God: The Failed Hypothesis; How Science Shows That God Does Not Exist*. Amherst, NY: Prometheus, 2006.

Stump, J. B., ed. *Four Views on Creation, Evolution, and Intelligent Design*. Grand Rapids: Zondervan, 2017.

Sullivan, Laura L. *Charles Darwin: Groundbreaking Naturalist and Evolutionary Theorist*. Minneapolis: Abdo, 2016.

Tao Te Ching. Translated by Gia-Fu Feng and Jane English. New York: Vintage, 1997.

The Udana. Translated by Dawsonne M. Strong. London: Luzac, 1902.

The Upanishads. Edited and translated by Valerie J. Roebuck. New York: Penguin, 2003.

Van Doren, Charles. *A History of Knowledge: Past, Present, and Future*. New York: Ballantine, 1991.

The Vedas. Edited by Jon William Fergus. Translated by Ralph T. H. Griffith and Arthur Berridale Keith. North Charleston, SC: Kshetra/CreativeSpace, 2017.

Watson, J. D., and F. H. Crick. "A Structure for Deoxyribose Nucleic Acids." *Nature* 171 (1953) 737–38.

Weinberg, Steven. *To Explain the World*. New York: HarperCollins, 2015.

Young, William A. *The World's Religions: Worldviews and Contemporary Issues*. 4th ed. Boston: Pearson, 2013.

Subject and Author Index

Printed in the USA
CPSIA information can be obtained
at www.ICGtesting.com
LVHW011515211023
761548LV00002B/36

9 781666 702385